Learning and Coordination

International Series on

MICROPROCESSOR-BASED AND INTELLIGENT SYSTEMS ENGINEERING

VOLUME 13

Learning and Coordination

Enhancing Agent Performance through Distributed Decision Making

by

STEVEN H. KIM
Lightwell Inc.,
Charlottesville, Virginia, U.S.A.

SPRINGER-SCIENCE+BUSINESS MEDIA, B.V.

Library of Congress Cataloging-in-Publication Data

Kim, Steven H.
Learning and coordination : enhancing agent performance through distributed decision making / by Steven H. Kim.
p. cm. -- (International series on microprocessor-based and intelligent systems engineering ; v. 13)
Includes index.
ISBN 978-94-010-4442-4 ISBN 978-94-011-1016-7 (eBook)
DOI 10.1007/978-94-011-1016-7
1. Neural networks (Computer science) 2. Machine learning.
3. Artificial intelligence. I. Title. II. Series.
QA76.87.K56 1994
003'.7--dc20 94-30344

ISBN 978-94-010-4442-4

Printed on acid-free paper

To the memory of my grandfather

Yong Joo Kim

Contents

Preface

Learning is a hallmark of intelligence. As such, the subject fascinates us all, whether we are participants at a technical conference, students of intelligent systems, or guests at a cocktail party.

The psychologist might define learning as a relatively permanent change in behavior as a result of experience. But this perspective ignores the positive factor implicit in the term *learning*. In everyday parlance, we associate learning with an improvement in performance rather than neutral or even degraded behavior. In operational terms, then, we may define learning as the *enhancement of performance over time*.

When a system is a product of human fabrication rather than the natural environment, its performance is often specified in terms of a set of functional requirements. These requirements indicate the level of performance expected of the system. In an uncertain world subject to stochastic disturbances, the fulfillment of functional requirements is a probabilistic rather than deterministic affair. In fact all systems—whether of the learning variety or not— operate in an uncertain environment.

But there is a second aspect of uncertainty: that of the system itself. The components of the system, both in terms of hardware and software, may be subject to stochastic influences.

This book examines a number of models which exhibit learning behavior, ranging from the human brain to artificial systems. The stochastic nature and learning behavior of intelligent systems are explored.

Intelligent systems must often interact in the attainment of their goals. For such systems, the ability to coordinate their activities prudently may prove to be more instrumental in fulfilling objectives than their portfolio of individual capabilities. As with individual strategies, the formulation of effective policies for coordination relies as well on probabilistic arguments. The resulting models and tools can then be used as a basis for improving collective performance with experience.

Given the universality of the subjects – namely, learning behavior and coordinative strategies in uncertain environments – this book should be of interest to students and researchers in diverse disciplines. These fields range from all areas of engineering to the computing disciplines; from the life sciences to the physical sciences; and from the management arts to societal studies.

I would like to thank Elias Towe and Mark Novick for reviewing the manuscript. Much of the word processing and proofreading were ably performed by Tasha Seitz. The thoughtful comments from these contributors were invaluable in improving the readability of the book.

S.K.

Introduction and Framework 1

Learning systems are characterized by a number of common features. This chapter presents a unified framework for adaptive systems. The framework comprises conceptual and formal constructs based on a modular representation of functional elements. The overall architecture is interpreted in specific contexts relating to robotics and production.

Introduction

Learning refers to the modification of behavior patterns over time.[1] Such adaptive behavior is useful for dealing with unknown factors in current or future activities. More specifically, they are effective in applications where (a) predictive knowledge of the underlying process is incomplete, and (b) variations in inputs or environmental conditions may exceed the predetermined range of values, or may even assume wholly unexpected characteristics.[2]

An intelligent machine is one which exhibits a range of behaviors, each action corresponding to specific input stimuli. For a synthetic device, the mapping from input to output is determined by a human programmer or another artifact, and remains fixed unless respecified by external intervention.

More sophisticated systems, however, incorporate autonomous learning behavior. By learning from experience, an automated system can modify its own repertoire of behaviors over time. The advantages of an autonomous learning capability relate to the following factors:

- *Unavailable knowledge*. The knowledge required for effective performance is often unavailable. For example, the final thickness of a composite material emerging from a curing process depends on the history of temperature and pressure variables in ways which are still unclear to production engineers.

---

[1] Kim (1990), Ch. 1.

[2] This chapter is based on a number of previous publications: Kim (1988a; 1990, App. D).

- *Elusive knowledge*. Sometimes a person may possess expertise, but cannot explain how he knows. A mundane example is a driver who makes the decision, based on "gut" feeling, to overtake a slow car in the face of a truck approaching from the opposite direction. A precise mathematical model of this situation would require differential equations, a methodology unknown to most drivers and not always recommended in an emergency.

- *Transfer bottleneck*. Even when an expert knows, and knows how he knows, it is often a laborious task to extract and codify the knowledge for translation into software. Any number of examples can be found from past efforts to develop expert systems in engineering, business, science, and other fields.

These difficulties underscore the advantages of developing autonomously learning systems.

One way to categorize learning systems takes the form of explicit versus implicit representation of knowledge. Explicit knowledge representation refers to the encoding of domain knowledge in terms and concepts readily comprehensible to the user. Examples of explicit representation lie in the Eurisko program[3] for discovering new concepts, or the learning classifier approach.[4] Explicit knowledge may be declarative, as in the statement of facts, or procedural, as in the encoding of instructions.

In contrast, implicit techniques refer to the generation of appropriate output based on input stimuli, without the direct representation of knowledge for processing the information. The implicit approach is reflected in neural network models. In this scheme, information processing know-how is encoded in terms of multiplicative weights at nodes and on the arcs between pairs of nodes. A key advantage of the neural approach is that the system itself performs the bulk of the work for establishing the implicit knowledge. A second advantage lies in robustness, in that a neural network will degrade gracefully rather than catastrophically when its structure is modified in minor ways.

The explicit and implicit methodologies are described more fully below, following an integrated framework for modeling learning systems in areas such as robotics and production control.

General Framework for Learning Systems

A *learning system L* is an entity which operates in some *environment E*. The object *L* receives certain *input* stimuli Σ from its environment. The domain of Σ is a set of physical quantities such as pressure or temperature; its range is usually some subset of the real numbers.

---

[3] Lenat (1983).

[4] Holland (1975).

The general model[5] is depicted in Figure 1.1. The purpose of the system is stipulated in terms of a set of functional requirements.[6] The functional requirements define the set F of acceptable final states. The system L, as an adaptive unit, must incorporate mechanisms which specify its own behavior patterns. Hence L encompasses the input transition function M and the output function N, as well as the set F of final states plus the other internal states of the kernel subsystem K.

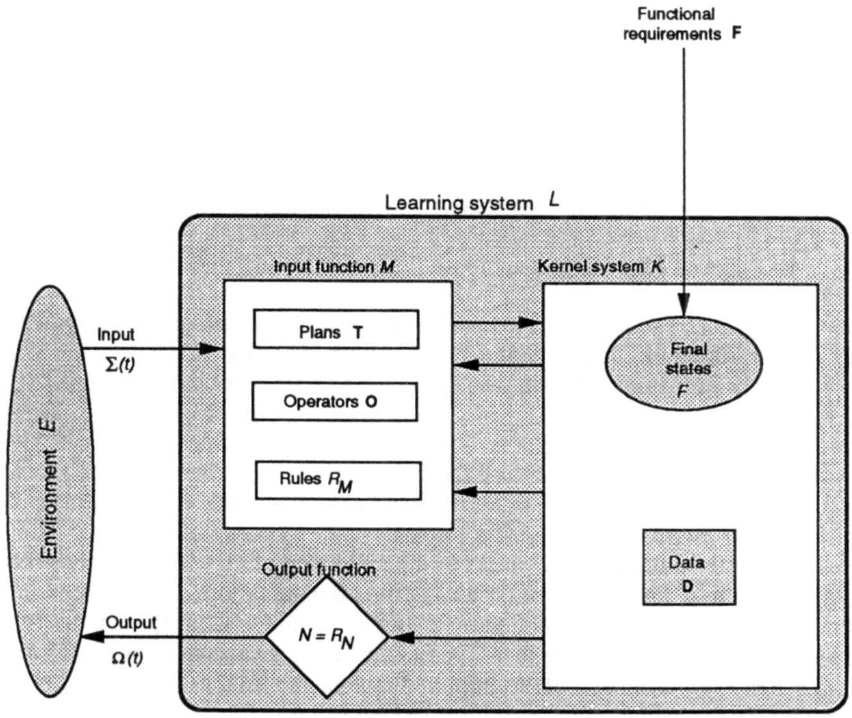

Figure 1.1 General framework for self-learning systems

The kernel K consists of the set F of final states and some data **D**. The terminal cluster F is specified by the functional requirements **F**, while the data **D** is determined by the messages from the input function $\boldsymbol{M}$.

The main component of the input transition function $\boldsymbol{M}$ is the set of input transition rules R_M. The function $\boldsymbol{M}$ also encompasses the meta-level operators $\boldsymbol{O}$ and the plans $\boldsymbol{T}$ of the adaptive system.

---

5 Kim (1988b).

6 Kim (1991); Kim and Ooi (1991); Kim and Suh (1987); Knight and Kim (1991).

The output function N is specified by the rules R_N which determine the external behavior of the system. These ideas may be encapsulated in the following algebraic model:

Definition. The structure $L = \langle \mathbf{S}, \mathbf{F}, \Sigma, \Omega, M, N, F \rangle$ defines a finite deterministic learning automaton L if and only if:

1. **S**, called the state space of L, is a finite nonempty set with n > 1 elements. Every element of **L** represents an internal state of L.

2. **F**, called the *functional requirement*, is a nonempty set of symbols. Each symbol specifies a goal state of L.

3. Σ is a set of input symbols.

4. Ω is a set of output symbols.

5. The function $M : \mathbf{S} \times \Sigma \rightarrow \mathbf{S}$, is called the *complete transition* (or next state) function. If $s \in \mathbf{S}$ and $x \in \Sigma^*$, then $M(s,x) \in \mathbf{S}$ denotes the state that the automaton L goes to when L, while in state s, receives input x.

6. The function $N : \mathbf{S} \times \Sigma^* \rightarrow \Omega^*$ is called the *output function.*

7. F is the set of final states defined by the functional requirements **F**.

In the definition above, Σ^* refers to the collection of all concatenations (ordered sets) of symbols in Σ; and similarly for Ω^* as the concatenations of symbols in Ω. These concepts may be further illustrated by a simple example. Consider a robot whose task is to assemble an automotive engine. The functional requirements **F** might be encoded as the final states F in terms of a photographic image of the final assembly as well as physical characteristics such as low friction in the crankshaft. Input sensors such as vision and touch provide information Σ which may be stored as data **D**.

The transition function M may regulate activities such as the sequence of assembly operations. The rule base R_M may contain knowledge such as "If the piston rings are in place, initiate the piston-crankshaft assembly process", or "If a piston shows discoloration, check for structural integrity". The set of plans **T** determines how the rule base R_M may be improved by selecting appropriate operators from the operator base **O**. The output function N would contain knowledge such as the manipulator motions required to insert pistons into place.

The deterministic automaton L models a device which changes state in predictable fashion on the basis of a particular input. This situation ensues when the transition rules are functions that exhibit unique domain values (i.e., "conditions", "antecedents", or "detectors"). This implies, for example, that no two decision rules in the rule base R_M have the same configuration of detectors.

In contrast, when two decision rules can respond to the same input (i.e., they have the same detector), then the device is nondeterministic. Further, if the competing rules may be assigned specific probabilities of being selected, then the system is a probabilistic automaton.

Many real-world processes exhibit random characteristics. This phenomenon gives rise, for example, to problems with statistical quality assurance in manufacturing environments.[7]

In such stochastic domains, the automata-theoretic model may be generalized to include probability distributions over state transitions.[8] These models can be used as the basis for the determination of stochastic behavior such as mean transition times and rates of adaptation.

Application to Production-Rule Systems

This section discusses the application of the general adaptive framework to a specific approach to learning systems. In this approach the knowledge is represented explicitly as a set of rules in terms of triggering conditions and corresponding actions. A specific category of the production approach is found in learning classifier systems.

Structure

In a production rule system, the system L operates by utilizing a self-learning strategy or adaptive plan **T** in order to attain a set of functional requirements **F** (see Figure 1.2). In a composites processing application, **F** might have 2 components relating to density and thickness in a cured part.

The degree to which the functional requirements are met is indexed by a performance measure π . If a binary range for π is chosen, for example, then we might use $\pi = 1$ if all the functional requirements are met, and $\pi = 0$ otherwise.

The domain of action of the adaptive plan **T** is the set of *rules* **A**. **A** may be interpreted as the potential knowledge base of decision rules that L can employ. In general, the knowledge base **A** is *potential* rather than *actual*: at any time t, L may have only a small subset of the possible set of rules.

The adaptive plan **T** yields a sequence of selections from a set of *operators* **O**. In other words, **T** invokes particular operators o in **O** to modify structures in **A**. An example of an operator is the splicing and mixing of two rules in **A**; this operator, known as a *genetic algorithm*, is discussed further in a later section.

---

7 Deming (1982); etc.

8 Kim (1985).

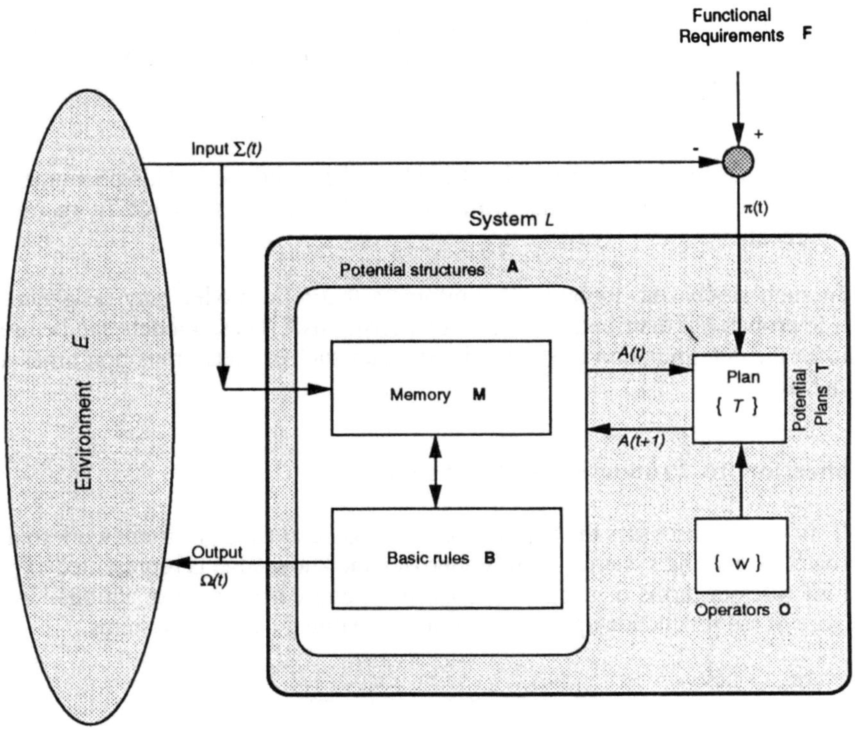

Figure 1.2 Framework for production-rule systems.

The set of general structures **A** may be partitioned into two subsets: the set of *basic structures* **B** and the *memory* **M**. We use the formalism **A** = ⟨ **B**, **M** ⟩ to indicate that **A** is the ordered pair consisting of **B** and **M**. For some basic rule *B* in **B**, let $B(t)$ denote the particular item applied at time *t* in order to fulfill the functional requirements **F**. At each cycle, the basic rules **B** observe the contents of memory **M** and may choose to generate a message. This message may be an output Ω directed to the environment or a memo *m* sent to augment the memory **M**.

Operation

In general, the adaptive plan **T** is a stochastic rather than deterministic process. Instead of a predetermined structure $A(t+1)$ obtained from $A(t)$ and $\Sigma(t)$, **T** assigns probabilities ⟨P1, P2, ...⟩ to the permissible set of rules ⟨A1, A2, ...⟩ and chooses the next rule to activate.

Let **P** be the set of permissible probability distributions over **A**. Then **T** first selects P(t+1) from **P**, followed by a selection of A(t+1) from **A**. For the deterministic case, P(t+1) assigns the single value 1 to a particular rule *A* in **A**.

In practice, the rule $A(t+1)$ is obtained from $A(t)$ by using an operator from the set $\mathbf{O} = \{ o: \mathbf{A} \rightarrow \mathbf{P} \}$. The adaptive plan is then given by the mapping $\mathbf{T}''$ which translates the input and rule set into the output. In other words, $\mathbf{T}'': \Sigma \times \mathbf{A} \rightarrow \Omega$.

Comparison of Adaptive Plans. A key requirement in using self-learning systems is to determine effective adaptive plans under differing environmental conditions. Let $\mathbf{T}'$ be the set of adaptive *plans* to be *compared* for use in the set of possible environments $\mathbf{E}$.

For a specific environment E in $\mathbf{E}$, the input Σ to the plan must contain some measure of the efficacy of $\mathbf{T}$ in fulfilling the functional requirements $\mathbf{F}$. In other words, a component of $\Sigma(t)$ must contain the payoff $\pi(\mathbf{A}(t))$, given by the mapping $\mathbf{A} \rightarrow$ *Reals*, which measures the degree to which the functional requirements are met in the environment E.

In the special case where the adaptive plan acts only on a knowledge of the reward, $\Sigma(t) = \pi_E(A(t))$. Plans which utilize information in addition to reward should perform at least as well as those which use only payoff.

To compare the efficiency of adaptive plans against each other, we must define some criterion k. Constructing such a criterion is a nontrivial task. For example, consider the following two plans: one performs superbly in some environments but poorly in others, while the other performs reasonably well in all cases. In this case, the selection of one plan over another should be guided by the particular application.[9]

Systems without Adaptive Capability. The formal framework discussed in the preceding sections may be tailored to particular domains. For example, it may be applied to intelligent manufacturing systems whose capabilities remain constant over time. Since learning is absent in such a system, the memory set $\mathbf{M}$ consists of the following: a singleton $\Sigma(t)$ at time t, and the empty set for the memo bank $\mathbf{m}$. $\Sigma(t)$ may, of course, still incorporate payoff information $\pi(t)$ which measures how well the functional requirements are satisfied.

The basic rule set $\mathbf{B}$ is static: it remains unchanged from one time instant to the next. Hence the "adaptive" plan $\mathbf{T}$ is a degenerate mapping from $\Sigma \times \mathbf{A} \rightarrow \mathbf{A}$. In other words, the rule at time t is given by:

$$\mathrm{A}(t) = \langle B, M(t) \rangle = \langle B, \Sigma(t) \rangle$$

[9] An example of such an application-dependent factor lies in the severity of consequences and the level of risk aversion involved (Kim, 1992).

Interpretation for Classifier Systems

Structure. A category of rule-based systems which has been studied extensively is that of *learning classifier systems.*[10] The classifier technique has been applied to a variety of simplified domains, from negotiating mazes[11] to controlling simulated pipelines.[12]

A classifier system is a program which learns rules called classifiers which define the system's response to some environment. The two central features of a classifier system are:

- Rule and Message System,
- Credit Assignment System.

A schematic diagram is given in Figure 1.3. A popular method of assigning credit and modifying the rule base is the genetic algorithm.

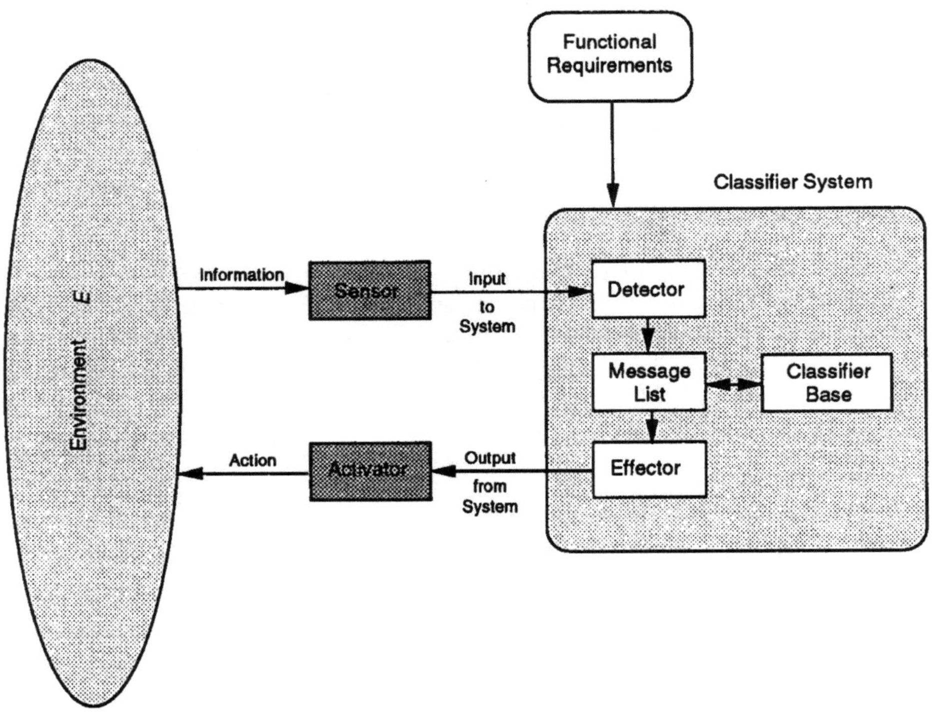

Figure 1.3 Schematic of a classifier system.

---

10 Holland (1975, 1986).

11 Holland and Reitman (1978).

12 Goldberg (1983).

The *detector* subsystem receives input concerning the state of the environment; the input information is decoded according to a prescribed *message* format. This message is placed in a common area called a *memo bank*, *blackboard*, or *message list*, together with other internal messages remaining from the previous cycles. Messages on the message list can stimulate rules called *classifiers*, stored in the *classifier* base.

An activated classifier may be selected to send a message to the message list for the next cycle. In addition, other messages may specify external action through a set of rules called *effectors*.

Operation. In conventional rule-based systems, the value of a rule is specified by the system developer. In a learning classifier system, however, the relative utility of each rule is a vital item of information which must be determined by the system itself.

Such learning is effected by the competitive economy in which classifiers bid for the right to respond to relevant messages. The highest bidders are awarded the contract; if they are successful in fulfilling their goals, then they are rewarded with an income which may be used for bidding at the next cycle. In such a competitive economy, the fittest rules survive and the unfit eventually disappear.

In the natural arena, species survive dramatic changes in the environment by mutation and the juxtaposition of different genes. Similarly, a classifier system provides for the generation of new rules through the mechanism known as the genetic algorithm.

This algorithm involves the generation of new rules through the mutation of old ones. One mechanism for mutation is the interchange of the components of two rules (analogous to chromosomes) after splicing at a random position. For example, consider two strings $S_1 = \langle B, A, ?, B, A \rangle$ and $S_2 = \langle A, B, B, A, ? \rangle$. If the random position selected is 2, then one offspring T_1 obtains the first 2 genes of S_1 and the last 3 genes of S_2; that is, $T_1 = \langle B, A, B, A, ? \rangle$. Conversely, the other offspring is $T_2 = \langle A, B, ?, B, A \rangle$.

Speed of learning. The classifier approach springs from the genetic mechanism underlying biological evolution. Since random mutations in conjunction with natural selection appear to be effective in the natural environment, the methodology is highly appealing. To date, however, progress in theoretical foundations has been limited. In addition, practical applications have been constrained by the large time scales required for proper training. Depending on the application, the classifier approach may require thousands or millions of trials before a system attains successful performance.

The classifier approach will not be examined at length in this book. Given the inherent affinity of the approach to biological systems, however, genetic algorithms richly deserve the current efforts of researchers around the globe.

Application to Neural Networks

This section interprets the general adaptive model in the context of neural nets, also known as the connectionist model. For this model, we may specialize the general framework to the neural model shown in Figure 1.4.

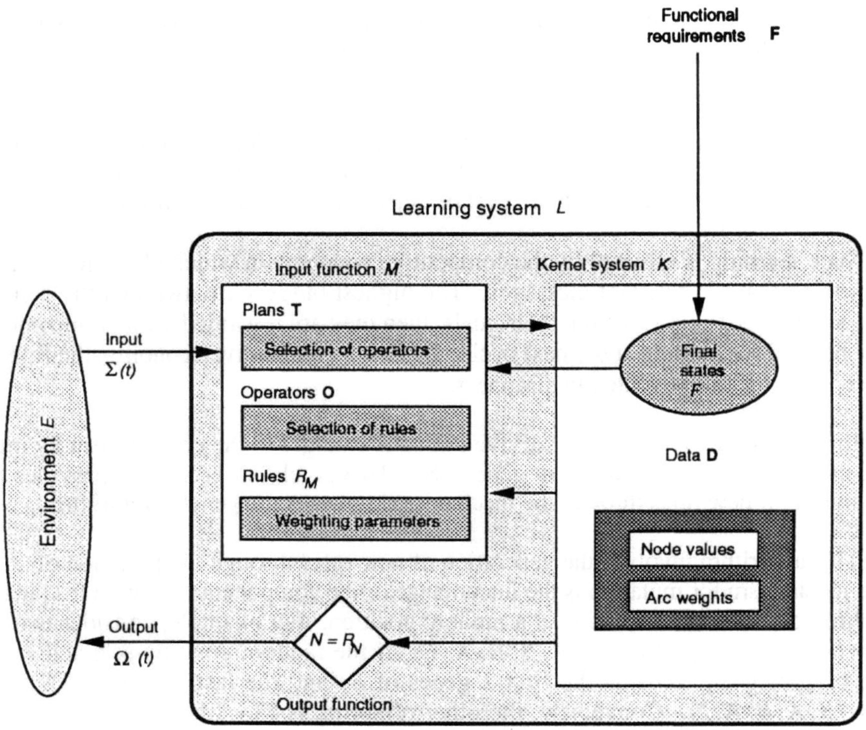

Figure 1.4 Framework applied to the neural network model.

In general, the connectionist approach does not store input data **D** in explicit form. Rather, the inputs are absorbed into the organizational structure of the system. More specifically, the input is transformed into a set of node values plus arc weights which depict the relationships among the nodes.

The transformation rules are procedures which specify how the data is to be converted as a function of the input and the current system state. These rules are defined by both the weighting strategies as well as the parameters used; an example is found in the "momentum" factors which specify how much influence the new input is to have on the existing state.

The operators define which rules to activate or modify. Current connectionist models tend to use a null operator set; that is, the initial rules for a system remain permanently active and are unchanged for the entire simulation. The plans pertain to the selection of a specific set of operators for a particular combination of system and environment.

The general connectionist model may itself be specialized for specific applications. For example, the feedforward connectionist model[13] is a neural net in which the lateral arc weights may be considered null (i.e., having zero value). The neural network approach will be examined at length in Chapter 2.

Explicit vs. Implicit Representation

The obvious advantage of explicit over implicit knowledge representation is the ready interpretation of the newly acquired knowledge. Ideally, the system should be able to inform the user of its findings, not merely transform a set of inputs into effective outputs.

Unfortunately, existing connectionist techniques do not allow for this facility. We might, however, take a cue from the human brain (as well as levels of computer languages, from machine language to high level specifications.) At lower levels of representation in the human brain, data is encoded in the form of voltage differences and structural changes, both mediated by chemical action. Yet, at some level of awareness, we are conscious of aggregate concepts (e.g., systems), relationships, (e.g., bondings), and abstractions (e.g., freedom).[14] Over the longer term, we must find ways to effectively couple the implicit and explicit forms of representing knowledge.[15]

Software Structure

Since the software organization must implement the generic functionality of a learning system, its architecture will mirror the functional organization given in Figure 1.1. To illustrate, consider the software structure of a classifier system. A central design issue pertains to the structure of the set **B** of basic rules. Each decision rule or *generalized chromosome* in the basic set **B** must consist of a detector, effector, and strength. The *strength* of the chromosome is a measure of its utility. A chromosome having a high strength value should be retained, while one with a low value is a candidate for elimination.

---

[13] Pao (1988).

[14] In this vein, one interesting approach is the iterative generation of aggregated concepts as implemented in the BACON program (Langley *et al.* 1983). This program seeks patterns among sets of data pertaining to different parameters or variables. The program has prior knowledge of the candidate algebraic relationships, such as the multiplication or division of variables, and seeks out the actual relationships consistent with the data sets. This technique has been used to "rediscover" a number of algebraic relationships in chemistry and physics, such as Ohm's Law which specifies voltage as the product of current and resistance.

[15] Shibazaki and Kim (1991).

The *detector* identifies the conditions under which the chromosome may "fire" or be activated. The *effector* specifies what set of actions is to be taken when the chromosome is activated. The components of each detector or effector may be called *generalized genes.* Each gene may take one of a set of values; each such value is called a *generalized allele.*

These concepts may be readily expressed in Pascal-like statements. Let detectorAlleleType be the set of allele values which may be assigned to each detector gene; the number of such genes in a detector is specified by the constant MAXDETECTOR. Similarly, let effectorAlleleType be the set of alleles for MAXEFFECTOR genes in the effector. Then the detectors and effectors can be prescribed to be instances of the following types:

```
detectorGeneType : detectorAlleleType;
effectorGeneType : effectorAlleleType;
detectorType = Array[1. . MAXDETECTOR] Of detectorGeneType;
effectorType = Array[1 . . MAXEFFECTOR] Of effectorGeneType;
```

For example, each element of the detectorType array is a gene whose value is taken from the pool of values in detectorAlleleType.

Using these definitions, we may specify each chromosome as:

```
chromosomeType = Record
  strength : Real;
  detector : detectorType;
  effector : effectorType;
        End;
```

In particular, strength is a variable which can accept a real value. A representative range for this parameter is the set of real numbers from a low of 0 to a high of 1.

If there are MAXCHROMOSOME number of chromosomes in the learning system, its genetic library is of the following type:

```
chromosomeBaseType = Array[1 .. MAXCHROMOSOME] Of chromosomeType;
```

This final structure corresponds to the set **B** of basic structures depicted in Figure 1.2. These concepts are clarified further through examples in the next section.

Illustrative Applications

Mobile Robot

The concepts introduced in the preceding section may be clarified through application to a simple context pertaining to a mobile robot. Suppose that a robot R is placed in an amusement park with the purpose of picking up litter such as empty soda cans; the robot might just as easily seek out exceptional conditions in an automated factory. We assume the following:

1. *R* can distinguish eight directions in its immediate neighborhood: north (*N*), northeast (*NE*), east (*E*), southeast (*SE*), south (*S*), southwest (*SW*), west (*W*), and northwest (*NW*). Each of these regions in *R*'s immediate neighborhood is called a *cell.*
2. *R* can recognize three types of objects in any cell: open space (*S*), obstruction (*O*), and litter (*L*). An example of an obstruction is a person, a cart, or a pillar; an example of litter is a stray popcorn container.
3. Each cell contains exactly one object: *S*, *O*, or *L*.
4. At each cycle, *R* moves into an adjacent cell.
5. *R* does not move into any cell containing an obstruction.
6. *R* succeeds when it moves into a cell containing litter (which it recovers during the move).

In this scenario, the learning system is the robot *R*, having the functional requirement of gathering litter. The detector component of each chromosome consists of eight generalized genes corresponding to the eight directions; each gene can be expressed as one of three alleles {space, obstruction, or litter}. The effector for each chromosome consists of a single gene whose alleles are the eight directions.

An example of a chromosome might be

0.5;
S, ?, O, S, S, S, L, S ;
W

The interpretation is: "If there is a space (*S*) in the north cell, anything (?) in the northeast, an obstruction (*O*) to the east, litter (*L*) to the west, and space (*S*) everywhere else, then go west (*W*)." The strength of this rule is 0.5 on a scale of 0 to 1.

The robot may begin with a random collection of such chromosomes. By rewarding chromosomes which lead to successful behavior and penalizing others, *R* will eventually accumulate an effective set of rules.

This is particularly true when the environment exhibits some orderly – rather than purely random – behavior. For example, suppose that litter is to be found to the north of every pair of blocking objects. Then the robot should ultimately behave in a way that takes account of this regularity in the environment.

Production Process Control

This section illustrates the adaptive framework by describing it in the context of production control applications such as composites fabrication. Composites are used in a wide variety of applications; an illustrative application is found in advanced aircraft bodies where they are used for their superior strength-to-weight ratios.

Although standards exist for many composites, their characteristics vary from lot to lot even when produced by a single manufacturer. Because the chemical processes and curing characteristics are not fully understood, final part quality may differ significantly due to factors such as the existence of voids and differing chemical structures in the cured part.

A straightforward control strategy is to rely on simple open-loop processes. However, a more effective methodology is to incorporate the following features: closed-loop control, real-time sensing, and learning techniques.[16]

Figure 1.5 depicts an advanced composites curing scheme. The process is designed to take material input in the form of fibers and resin, and to produce composites parts with specific functional requirements such as thickness H, fiber volume fraction V_f (i.e., volume of fibers as a fraction of the total part volume), or uniform resin density ρ. The functional requirements are to be met by design parameters such as the mold temperature T and pressure P. The penalty function is defined by the actual and desired values of the requirements. Its mathematical structure will depend on the particular application.

Discussion

We have examined a unifying framework for integrating and examining issues relating to learning systems. By addressing these topics in a coherent fashion, we may construct a basis for analyzing intelligent artifacts and synthesizing learning systems in diverse domains.

Intelligent systems should improve their performance over time. In applications of practical consequence, systems usually operate under probabilistic conditions. Such conditions can be characterized by random disturbances; or the information available to the system may be incomplete; or the consequences of potential actions might only be partially understood. In such a context, performance enhancement is tantamount to increasing the likelihood and degree of fufilling the functional requirements of the system.

Learning techniques may relate to the performance of individual systems as well as their ensembles. For individual decision-making agents, stochastic learning techniques may be applied. For multiagent systems, the proper coordination of individual actions might be of greater consequence than the optimization of individual behavior. In such cases, the methods of game theory and related methodologies may be applied to enhance the performance of the group as a whole.

---

[16] Egilmez and Kim (1989a, 1989b, 1990, 1991a, 1990b); Hagler and Kim (1987); Kutluoglu and Kim (1989).

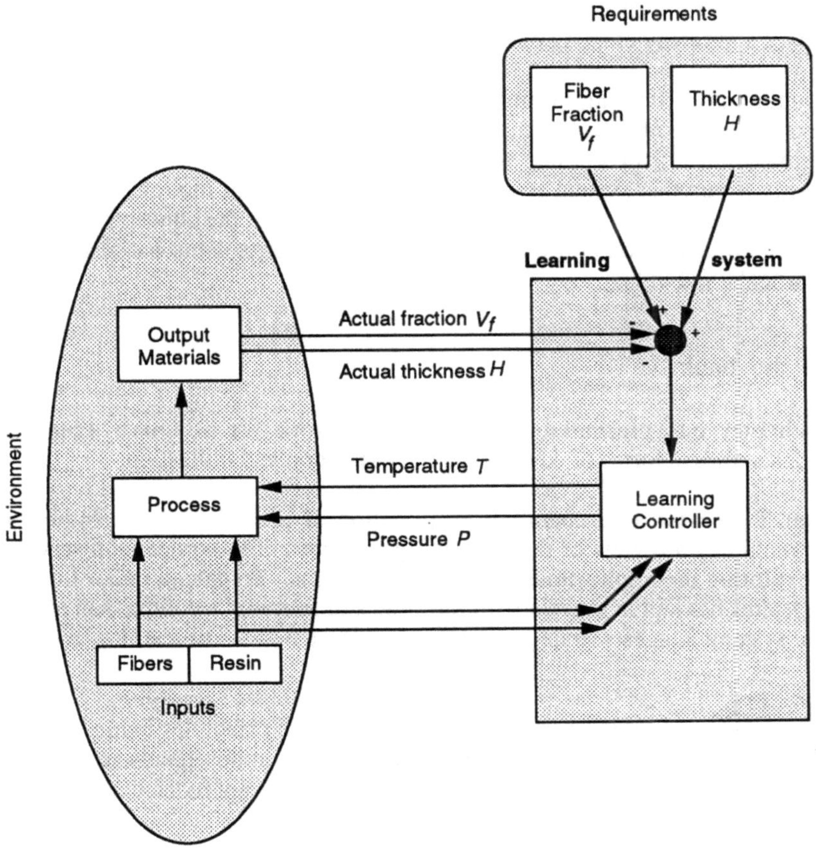

Figure 1.5 Some critical parameters for composites fabrication.

Competition in biological systems occurs not only at the level of organisms, but at the lower plane of organs and even cells. At the level of organs, a blind person may well develop more acute hearing than a sighted individual; this enhanced ability can compensate for the deprivation of the visual sensory mode.

We may identify instances of competition even within the visual apparatus. Some individuals develop a "lazy eye" which exhibits poor visual acuity; this results from over-reliance on the other, dominant, eye. At the level of tissues, another example of competition for bodily resources lies in muscle growth: muscles enlarge or atropy, depending on the amount of use.

At the cellular level, the competitive arena can mete out disablement and even death. A neuron which is unused will wither and suffer from weakened synapses; at the extreme, the cell may die as part of the larger developmental processes which fine-tune the brain for survival in an uncertain environment. On the other hand, an active neuron will generate new synapses and fortify existing ones.

Computational processing in biological systems exhibits modularization as well as distribution. More specifically, certain functions such as language may be allocated to one set of regions, sensory processing to another, motor activities to a third, and so on. Consequently a nervous system is a collection of interacting modules, each of which enjoys partial autonomy and decision capability. As a result, the concepts and principles of multiagent collaboration apply not only to disparate organisms but also to modules within a particular individual.

Scope of the book

This chapter has presented a general framework for adaptive agents which accommodates diverse forms of knowledge representation and utilization.

Chapter 2 presents a comparative examination of artificial and biological neural networks, including a review of the underlying concepts. The comparison serves as the basis for exploring the implications for artificial systems. To date, a major limitation of artificial systems has lain in the relatively slow rates of learning; a comparison of synthetic and natural systems suggests strategies for accelerating the speed of learning in engineered devices.

Chapter 3 examines a number of rigorous models and principles behind cooperative behavior. The types of potential interactions among autonomous agents are discussed, as well as the limits to their collective behavior. A probabilistic approach to credit allocation is described, and its advantages examined.

The subsequent chapter presents a case study in multiagent coordination. The scenario is the effective deployment of robotic agents in a service function, whether in the form of security agents in an office building or shuttles on a factory floor. A number of deployment strategies are examined, including the probabilistic approach to credit allocation described in Chapter 3.

Finally, Chapter 5 offers some concluding remarks. This part summarizes the results of the previous chapters and explores a number of implications for the future.

The appendix describes dynamic models in the field of statistical physics. The statistical methodology, including its application to simulated annealing, serves as the basis for a number of approaches for computational convergence in neural networks.

The statistical methodology may prove instrumental in future work in modeling the collective action of numerous independent agents. When agents act en masse, their individual contributions will often combine in additive fashion. In that case, random fluctuations will tend to cancel, leaving only mean effects to exert their impact in collective activity.

References

Deming, W. E. *Quality, Productivity and Competitive Position.* Cambridge, MA: M.I.T. Center for Advanced Engineering Studies, 1982.

Egilmez, K. and S.H. Kim, "Design and Control of Manufacturing Systems: An Integrated Approach Based on Modal Logic and Zonecharts". *Robotics and Computer-Integrated Manufacturing,* v. 6(3), 1989a: 209-228.

Egilmez, K. and S. H. Kim. "Zone Logic for Knowledge-Based Control: Formalization and Applications." *Proc. IEEE International Symp. on Intelligent Control,* Albany, NY, Sep. 1989b: 607-613.

Egilmez, K. and S. H. Kim. "A Logical Approach to Knowledge Based Control". *J. of Intelligent Manufacturing,* v. 1, 1990: 59-76.

Egilmez, K. and S. H. Kim. "Coordination of Multiple Agents through Modal State Logic and Influence Specifications; Part I, Models". *J. of Systems Engineering,* v. 1(1), 1991a: 2-12.

Egilmez, K. and S. H. Kim. "Coordination of Multiple Agents through Modal State Logic and Influence Specifications; Part II, Applications". *J. of Systems Engineering,* v. 1(2), 1991b: 77-93.

Goldberg, D. E. *Computer-aided Gas Pipeline Operation Using Genetic Algorithms and Rule Learning.* Ph.D. thesis, Univ. of Michigan, 1983.

Hagler, C.D. and S.H. Kim, "Information and its Effect on the Performance of a Robotic Assembly Process", *Proc. Symp. on Intelligent and Integrated Manufacturing: Analysis and Synthesis,* Boston, MA, Dec. 1987: 349-356.

Holland, J. H. *Adaptation in Natural and Artificial Systems.* Ann Arbor, MI: Univ. of Michigan Press, 1975.

Holland, J. H. "Escaping Brittleness: The Possibilities of General-Purpose Learning Algorithms Applied to Parallel Rule-Based Systems." In R. S. Michalski et al., eds., *Machine Learning: An Artificial Intelligence Approach,* v. II, Los Altos, CA: M. Kaufmann, 1986, pp. 593-624.

Holland, J. H. and J. S. Reitman. "Cognitive Systems Based on Adaptive Algorithms." In D. A. Waterman and F. Hayes-Roth, eds., *Pattern-Directed Inference Systems,* NY: Academic, 1978, pp. 313-329.

Kim, S. H. *Mathematical Foundations of Manufacturing Science: Theory and Implications.* Ph.D. Thesis, M.I.T., Cambridge, MA, 1985.

Kim, S.H., "Managing Complexity in Automated Systems: An Information Measure and Its Application", *Manufacturing Technology Review and North American Manufacturing Research Conference Proc.*, Bethlehem, PA, May 1987: 541-545.

Kim, S.H., "A Unified Framework for Self-Learning Systems". *Proc. Manufacturing International '88, v. III: Symp. on Manufacturing Systems*, Atlanta, GA, April 1988a: 165-170.

Kim, S.H., "An Automata-Theoretic Framework for Intelligent Systems". *Robotics and Computer-Integrated Manufacturing*, v. 5(1), 1988b: 43-51.

Kim, S. H. *Designing Intelligence: A Framework for Smart Systems*. New York: Oxford Univ. Press, 1990.

Kim, S.H. "Product Performance as a Unifying Theme in Concurrent Design – I: Concepts". *Robotics and Computer-Integrated Manufacturing*, v. 8(2), 1991: 121-126.

Kim, S. H. *Statistics and Decisions: An Introduction to Foundations*. New York: Van Nostrand Reinhold, 1992.

Kim, S.H. and J. Ooi. "Product Performance as a Unifying Theme in Concurrent Design – II: Software". *Robotics and Computer-Integrated Manufacturing*, v. 8(2), 1991: 127-134.

Kim, S.H. and N.P. Suh, "Mathematical Foundations for Manufacturing", *Trans. ASME/ J. Engineering for Industry*, v. 109(3), 1987: 213 – 218.

Knight, T.P. and S.H. Kim, "A Knowledge System for Integrated Product Design". *J. of Intelligent Manufacturing*, v. 2, 1991: 17-25.

Kutluoglu, S.F. and S.H. Kim. "Information Characteristics and their Impact on System Performance: Framework and Case Study in Distributed Production Control". *Robotics and Computer-Integrated Manufacturing*, v. 6(3), 1989: 199-208.

Langley, P., G. L. Bradshaw, and H. A. Simon. "Rediscovering Chemistry with the BACON System." In R. S. Michalski et al., eds., *Machine Learning: An Artificial Intelligence Approach*, Palo Alto, CA: Tioga, 1983, pp. 307-329.

Lenat, D. "EURISKO: A Program that Learns New Heuristics and Design Concepts: The Nature of Heuristics, III: Program Design and Results." *Artificial Intelligence*, v. 21(2), 1983, pp. 61-98.

Pao, Y. H. "A Connectionist-Net Approach to Autonomous Machine Learning of Effective Process Control Strategies." *Robotics and Computer-Integrated Manufacturing*, v. 4(3/4), 1988, pp. 633-642.

Parthasarathy, S. and S.H. Kim, "Temporal Reasoning for Manufacturing: A Hybrid Representation and Complexity Management". *Robotics and Computer-Integrated Manufacturing*, v. 6(1), 1989: 67-81.

Parthasarathy, S. and S.H. Kim, "Manufacturing Systems: Parallel System Models and Some Theoretical Results". *International J. of Computer Applications in Technology*, v. 3(4), 1990: 225-38.

Shibazaki, H. and S.H. Kim, "Learning Systems for Manufacturing Automation: Integrating Explicit and Implicit Knowledge". Presented at International Conference on the Manufacturing Science and Technology of the Future, Stockholm, Sweden, June 1989. Also in *Robotics and Computer-Integrated Manufacturing*, v. 9, 1992: .

Learning Speed in Neural Networks 2

Intelligent systems should improve of their own accord over time. In the natural sphere, a proven technique for self-learning takes the form of neural networks.

In contrast to biological structures, however, artificial neural nets are saddled by long training requirements. A closer examination of biological systems highlights their differences from artificial networks. A hybrid paradigm involving both connectionism and hierarchical structures is presented, and its advantages examined in terms of speed of learning as well as other features.

Introduction

Intelligent systems should have the ability to enhance their performance based on past experience (Kim, 1990). This is a reasonable expectation whether the system is a machine tool on the production floor, a mobile sentry on a battlefield, or a switching center in a communication network.

A promising approach is that of the neural net, modeled after the neural structures of biological systems (Anderson and Rosenfeld, 1988; Hebb, 1949; Rosenblatt, 1962; Rumelhart et al., 1986). Networks offer many advantages including self-learning, graceful degradation, and other features. To date, however, they have been burdened with a major limitation: excessively long training periods[1]. Thousands and sometimes millions of trials are often required for satisfactory performance in various tasks. The time and effort required for training have hindered their widespread application to practical domains (Shibazaki and Kim, 1991).

[1] Onerous training seems to be an endemic problem in many learning systems. Long training periods are required not only for the neural net associated with individual learning, but the genetic approach to societal evolution (Gefenstette, 1985; Goldberg, 1983; Holland, 1975, 1986; Holland and Reitman, 1978). The same limitation hobbles learning activities using high-level representations. The program called Eurisko, for instance, has achieved practical goals such as a patentable electronic device and winning strategies in simulated warfare; but the program often required months of intense computational effort to achieve its results (Lenat, 1983).

It appears that the required training must be reduced dramatically, by orders of magnitude, before neural nets are likely to fulfill their full promise. Performance improvement of such magnitude will not likely materialize from a simple tweaking of algorithms or their parameters. A more drastic re-evaluation and improvement in technique are indicated.

Certain approaches to the speed-up of learning show promise. One of these relates to metal-level knowledge concerning the training procedure. For instance, changes in parameters such as weighting factors can be accelerated or reduced depending on the direction and magnitude of performance enhancement.

Another approach is a more sophisticated modeling of biological systems. In particular, we can incorporate a large-scale organizing structure atop the microlevel model of neural processing units.

To this end, we first examine the salient aspects of knowledge representation and processing in biological systems. To ground the discussion in tangible matters, the presentation will focus on the visual apparatus of primates.

The implications for synthetic networks are then discussed. A hybrid organization is proposed consisting of a hierarchical structure superimposed on the neural net. Finally, the advantages of this hybrid structure are discussed.

Neural nets

Neural nets have been described as a model of biological neural systems. Both types of systems are characterized by learning capability, the ability to improve performance over time. A closely related feature is that of generalization, relating to the recognition of new objects which are similar but not identical to previous ones. An additional feature is that of graceful degradation: the network fails gracefully rather than catastrophically when it suffers partial damage.

Neural networks seem sufficiently versatile to allow for near-optimal performance in simple behaviors. A study was performed involving the movement of a single polychaete worm into and out of a tube. It had to forage for food but could fall prey to a predator in the process. The performance measure was an index of Darwinian fitness for an individual based on its probability of survival and reproduction at the end of a trial. The nets incorporated four modules relating to motor function, sensing, energy reserves, and a biological clock (to indicate the end of a feeding cycle). In the context of neural networks, *ontogeny* refers to the change in connections for a single subject based on its experiences; on the other hand, *evolution* of the network pertains to natural selection among a collection of connective patterns. The study found that the best nets for individuals possessed fitness values of 85% to 99% of the performance resulting from the evolutionarily optimal net (Mangel, 1990). In this application, at least, learning within a single lifespan can approach that of intergenerational improvement.

Unlike biological systems, however, artificial neural nets require inordinate amounts of training. A person may learn to discriminate between a block and a cylinder after a single observation; but a net will require numerous trials. What accounts for this difference in learning rates? The short answer is that biological networks are more sophisticated, both in representing and in manipulating knowledge, than their upstart cousins made of semiconductors.

Learning Rate as a Performance Metric

A common performance characteristic for adaptive systems takes the form of learning speed. This parameter is often translated into the metric of the number of trials needed to fulfill functional requirements. The speed of learning, as a sole performance measure for learning algorithms, has its limitations. A sample limitation is the entrapment problem which lurks behind any nonlinear optimization problem: namely, settling into a local rather than global optimum.

In a more general context, entrapment can be viewed as a special case of the breadth-versus-depth problem. In particular, a learning algorithm which is "good enough" may be selected without taking into account features of other algorithms which could prove more robust or general. This problem arises in many domains. One such example is the pole balancing problem, where a cart on a rail can be moved to the left or right; it must support a pole that must be kept from falling in either direction (Samut and Cribb, 1990).

Types of Learning

Learning can be of two types, depending on the openness of the procedure. The goal of *closed* form learning is to associate a predetermined set of inputs with outputs. In contrast, the goal of *open* form learning is to discover previously unknown aspects of the available data.

- ***Closed form***. The object of this type of learning is to associate certain classes of inputs with corresponding outputs. Once a net has learned to perform this task properly, it can assign a particular output vector to any input vector from the input space. An example of this type of function lies in the form of the pattern associator.

  This type of activity is also called *classification* or *associative* learning. The training phase is called *supervised* training, since the net is presented with the "correct" response to each input.

- ***Open form***. The purpose of this type of learning is to discern common or interesting aspects relating to a set of data. There is no prior specification of "correct" or "incorrect" responses. The task here is comparable to the process of exploratory data analysis in the field of statistics. A net might learn, for instance, that the use of one material versus another enhances the reliability of an industrial product; or that economic activity peaks an average of 4 months after a similar rise in the stock market.

Since the goal here is to detect common or regular features in the data, the process is also called *feature extraction* or *regularity discovery*.

A neural net consists of a set of objects interacting with each other as well as with the environment (see Figure 2.1). Each object represents a set of functions or computations, as shown in Figure 2.2. These functions and the role of linkages among the objects are examined in detail below.

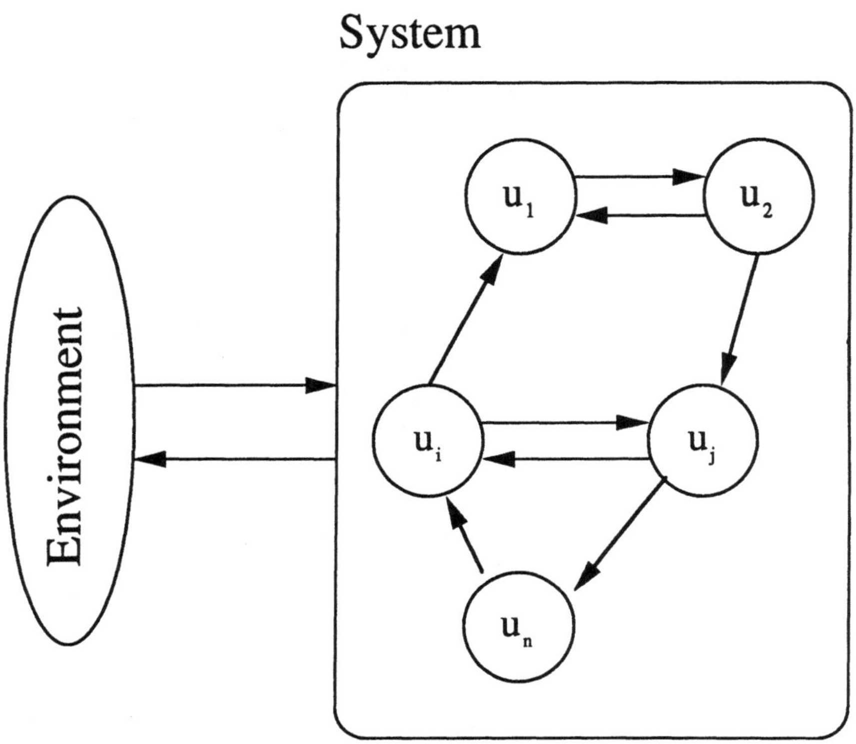

Figure 2.1 A neural net in interaction with its environment. The net consists of processing units or modules which communicate with each other and, as a whole, with the environment.

Activation Rules

Every unit or cell in a net takes some input and updates its internal state at each computational stage. Let $v_i(t)$ denote the input at time t to cell i:

$$v_i(t) = \Sigma_j \; w_{ij}(t) \; o_j(t)$$

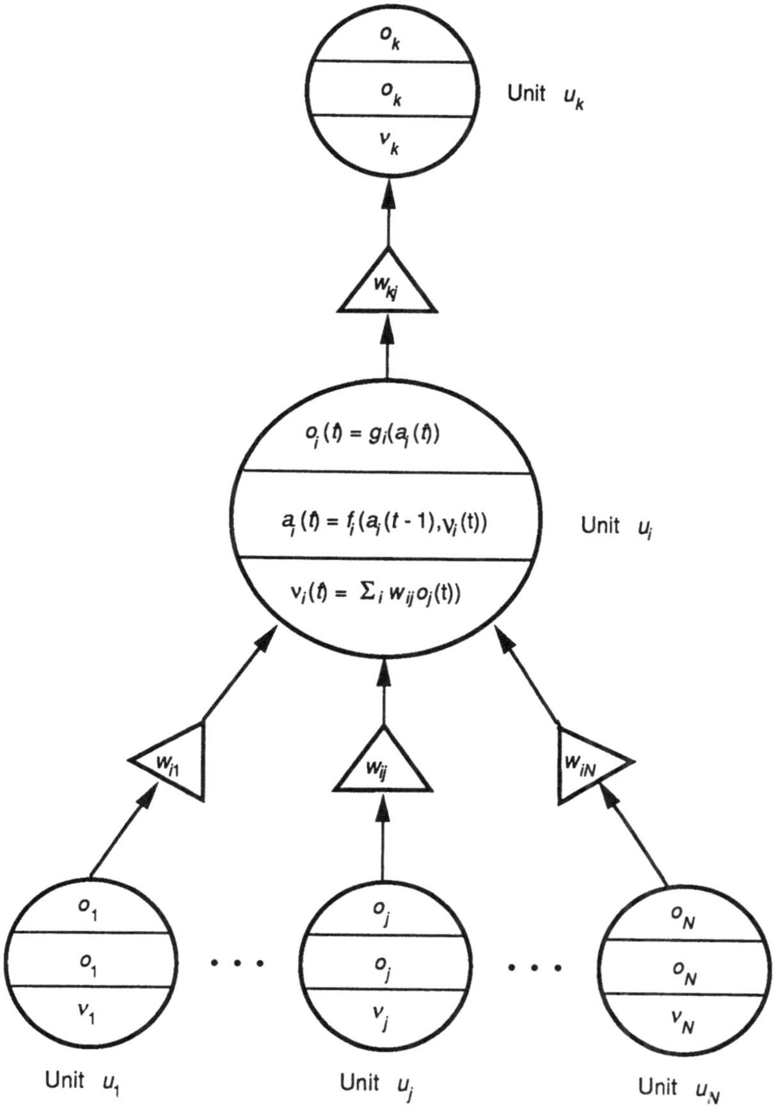

Figure 2.2 Computations for a processing unit of a neural network

Here $w_{ij}(t)$ is the weight between units u_i and u_j at time (t), while $o_j(t)$ is the output from unit j.

We let $v(t) \equiv (v_1(t), \ldots, v_n(t))^T$ denote the vector of input values, $o(t) \equiv (o_1(t), \ldots, o_n(t))^T$ the vector of output values, and $W=[w_{ij}]$ the matrix of weights. The superscript "T" denotes the transpose operation. Then we can write

$$v(t) \equiv W(t)\ o(t)$$

The activation for unit u_i at the next time step is a function of its extant activation $a_i(t)$ and the net input $v_i(t)$:

$$a_i(t+1) = f(a_i(t), v_i(t))$$

The transformation f is called the *activation rule* or *activation function*. This rule may be linear or nonlinear, and deterministic or probabilistic.

Output Rules

The output function g maps the activation level $a_i(t)$ at time t into some output value $o_i(t+1)$. In algebraic form, we can write

$$o_i(t+1) = g(a_i(t)) = g(f[a_i(t-1), v_i(t-1)])$$

The second equality follows from the fact that

$$a_i(t+1) = f(a_i(t), v_i(t))$$

In general, the output function g is taken to be the identity function. Since the output o_i is simply equal to the current activation level a_i, the identity "transformation" is assumed to require no time lag. Hence we can state

$$o_i(t+1) = g(a_i(t)) = a_i(t)$$

In terms of vector rotation, we can write o(t+1) = a(t).

To summarize, the output rule g is usually taken to be the identity function and regarded as requiring no time. From this perspective, g is the *null transformation*. In symbolic form, we can express the composition of functions g and f as simply f; that is, $g \circ f = f$.

The next section presents a number of activation rules. Unless otherwise stated, the output rule g will be taken to be the null tranformation. Hence the output simply equals the activation; in algebraic form, o(t) = a(t).

Examples of Activation Rules

Simple linear rule. The simplest type of activation rule f is the identity transformation. Here the new activation $a_i(t+1)$ equals the input $\nu_i(t)$. In algebraic form:

$$a_i(t+1) = f(a_i(t), \nu_i(t)) = \nu_i(t)$$

We note that the activation rule f is simply an identity function; but the "tranformation" is usually regarded as requiring one time step. The rationale is that replacing the old activation value $a_i(t)$ with the new value $a_i(t+1)$ requires a nontrivial time interval.

We can express the new activation vector in terms of the old output vector o(t) as follows:

$$a(t+1) = \nu(t) = W(t)\ o(t)$$

The second equality follows from the definition of the net input vector ν.

A neural net is usually regarded as consisting of three groups of cells: the input layer, hidden layer, and output layer. For the linear activation rule, however, the number of layers is inconsequential in the following sense. The capability of a net using multiple layers is *identical* to that using a single layer when the activation rule is a *simple linear function*. This result can be demonstrated as follows.

Let W denote the matrix of weights for a net after reaching steady state. In other words, $W(t) \rightarrow W$ for some sufficiently large time t. When the output rule g is the null transformation, the output vector simply equals the activation vector: $o(t) = a(t)$. Hence we can write

$$a(t+1) = W\ o(t) = W\ a(t)$$

For $t = 0$, we have $a(1) = W\ a(0)$. At the next step, $a(2) = W\ a(1) = W^2\ a(0)$. More generally, the activation at step k is given by $a(k) = W^k\ a(0)$.

Let V_k denote the matrix W^k. Then for any activation level a(k) reached in k time steps using matrix W^k, there is some other matrix V_k which could have attained the same result in one step. Consequently, the power of a multilayer net is identical to that of a single layer when the activation rule is a simple linear function.

Step function rule. The limitations of linear systems can be circumvented by introducing nonlinear transformations. Perhaps the simplest of these is the unit step function $u(\theta,\nu)$ defined as follows:

$$u(\theta,\nu) \equiv \begin{cases} 0 : \nu < \theta \\ 1 : \nu \geq \theta \end{cases}$$

The parameter θ is the threshold below which the output is nil, and otherwise the output is unity. This function is depicted in Figure 2.3.

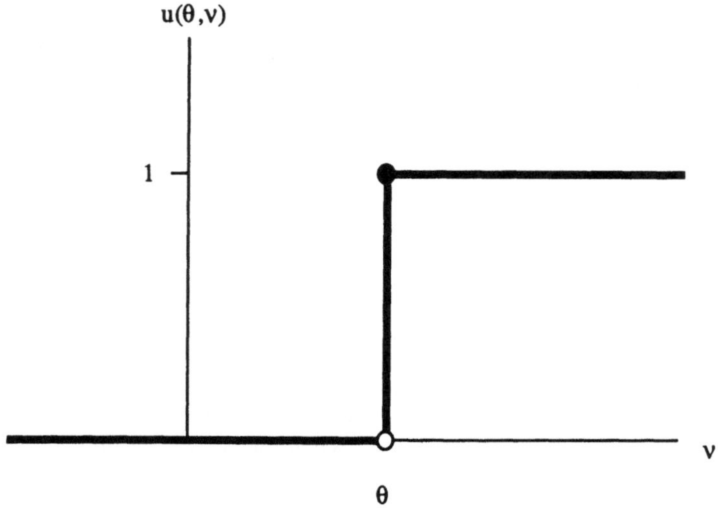

Figure 2.3 Unit step function.

As indicated previously, a nonlinear system has greater capability than a linear system. A simple demonstration of this superiority lies in the implementation of the *exclusive or* function, also called the *XOR* function. Given two inputs, each of which may be 0 or 1, the output of this function is 1 if precisely one of the inputs is 1; and the output is 0 otherwise.

A neural net to implement the XOR function is shown in Figure 2.4. Each cell implements the unit step function $u(0.5, \nu)$ with a threshold of 0.5. Let *p* and *q* be the inputs, and *r* the output. The 4 possible combinations of inputs – and the corresponding output – are given in the table to the right of the figure.

For example, the second entry in the table indicates that inputs of $p = 0$ and $q = 1$ result in an output of $r = 1$. More specifically, an input of $\nu_1 = p = 0$ to cell 1 results in activation and output values of 0:

$$o_1 = a_1 = 0$$

On the other hand, the input of $\nu_2 = q = 1$ to cell 2 yields activation and output values of 1:

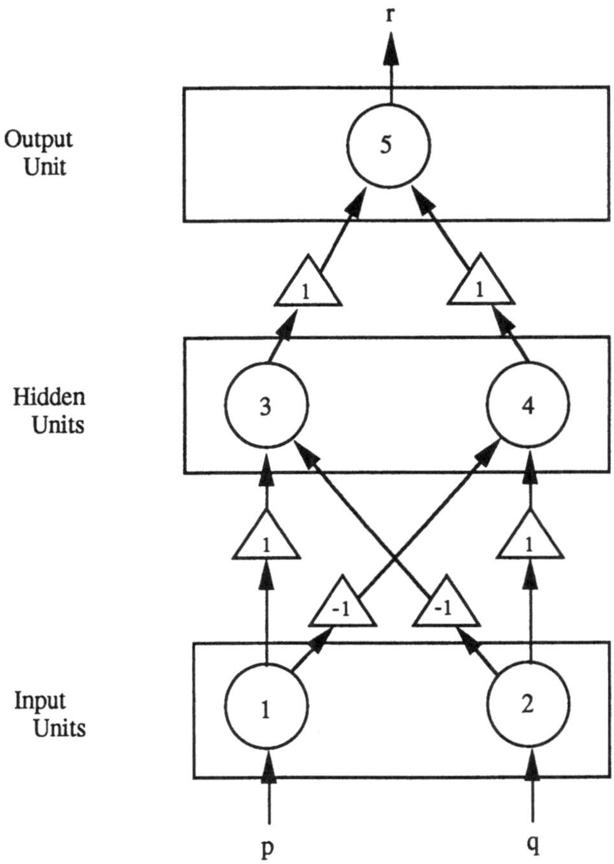

p	*q*	*r*
0	0	0
0	1	1
1	0	1
1	1	0

Figure 2.4 Implementation of the "exclusive or" function using the unit step function $u(\theta,\nu)$ with threshold $\theta = 0.5$.

$$o_2 = a_2 = 1$$

The inputs to cells 3 and 4 are given by

$$\nu_3 = w_{31}o_1 + w_{32}o_2 = 1(0) + (-1)1 = -1$$

$$\nu_4 = w_{41}o_1 + w_{42}o_2 = -1(0) + (1)1 = 1$$

Since the input to cell 3 is -1, the unit step function will yield an activation value of 0. In other words

$$o_3 = a_3 = 0$$

On the other hand, the input of 1 to cell 4 will yield an activation of unity:

$$o_4 = a_4 = 1$$

The input to cell 5 is

$$v_5 = w_{53}o_3 + w_{54}o_4 = 1(0) + 1(1) = 1$$

which is the desired result. The reasoning for other combinations of input values for *p* and *q* is similar.

Saturating linear rule. Another activation rule is the truncated linear function. The output equals the input for the range [–1, 1]; otherwise the output value is the sign of the input. The function is depicted in Figure 2.5.

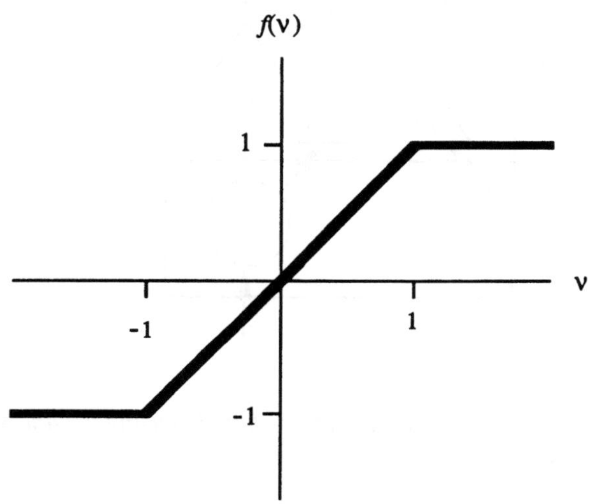

Figure 2.5 Saturating linear rule

Sigmoid rule. A popular type of activation rule is the sigmoid function. A *sigmoid* function is any S-shaped curve. An example lies in the function

$$f(\theta,T,v) = \frac{1}{1 + e^{-(v-\theta)/T}} \qquad \text{(Eq. 1)}$$

The parameter θ is a constant of location. When the input v coincides with θ, the activation value is one half:

$$f(\theta,T,\nu) = \frac{1}{1+e^0} = \frac{1}{2}$$

As ν approaches infinity, the value of *f* rises toward 1; as ν approaches negative infinity, *f* falls toward 0.

Figure 2.6 depicts the sigmoid function for the threshold value of θ = 10. Two curves are shown in the figure, depending on the parameter T. This parameter is a constant of scale which determines the "elongation" of the sigmoid curve. When T increases, the curve becomes more shallow. As T approaches infinity, *f* assumes the constant value of 0.5.

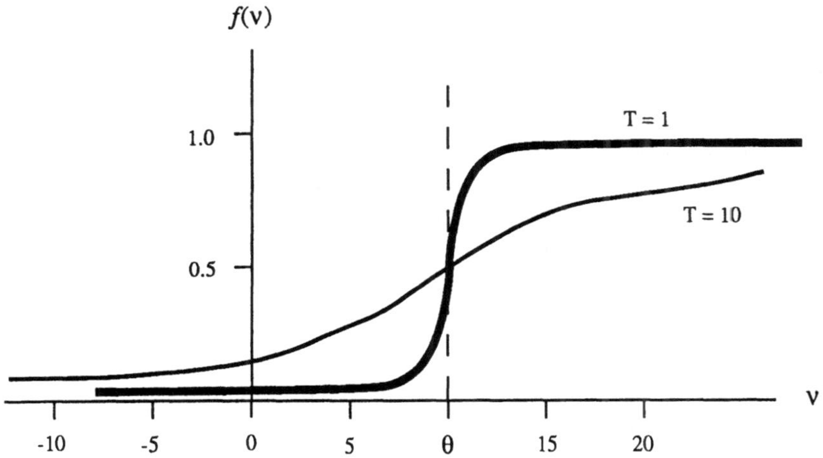

Figure 2.6 Sigmoid rule, for the threshold value θ = 10.

In contrast, consider the case where T is very small. Then any positive value of ν in excess of θ will be magnified many times in the exponential term in Eq. 1. The result will be

$$f(\theta,T,\nu) \approx \frac{1}{1+e^{-\infty}} = 1$$

Conversely, any finite value of ν below θ will cause the exponential term to explode, yielding $f(\theta,\nu) = 0$. In short, as $T \to 0$, the activation function $f(\theta,T,\nu)$ corresponds to the unit step function $u(\theta,\nu)$ with parameter θ.

The sigmoid function has been described here as a deterministic rule. It is simply a mapping from the set of reals into the interval [0,1]. In other words, $f: [-\infty, -\infty] \rightarrow [0,1]$.

The sigmoid function can also define a probabilistic transformation. To this we now turn.

Probabilistic sigmoid rule. Consider the case where the input can assume any real value but the output must be binary. In other words, the input is $\nu \in [-\infty, \infty]$ while the activation is $a \in \{0,1\}$.

The formula given in Eq. 1 can serve as the basis for a probabilistic function. In this context, the value of $f(\theta,\nu)$ indicates the *probability* that the output a assumes the value 1, rather than the value of a itself. Hence $a = 1$ with a probability of $f(\theta,T,\nu)$, and $a = 0$ with probability $1 - f(\theta,\nu)$.

The scaling parameter T is called the *temperature*. Formulas of the type given by Eq. 1 arise in the field of thermodynamics. The probability that a particular atom or molecule will enter one of two possible states is given by $f(\theta,T,\nu)$. The likelihood of the transition depends on the energy ν in excess of some threshold θ as well as the temperature T. Near absolute zero, $T \approx 0$ and any value of $\nu - \theta > 0$ will cause the transition to occur. Conversely, for high values of T, an atom is about as likely to enter one state as the other.

Multiplicative Links

Thus far, we have regarded the input to a cell as the straightforward weighted sum of outputs from various nodes. In algebraic terms, we have

$$\nu_i = \Sigma_j w_{ij} o_j$$

In biological systems, however, one neuron can empower or neutralize the function of another. This can be achieved in two ways:

1. *Dendritic innervation.* One neuron stimulates the dendrites on another, either positively or negatively. When the net input to a neuron exceeds some threshold value, it sends its own voltage pulse down its axon.

2. *Axonic innervation.* A neuron stimulates the axon of a second neuron, which innervates a third neuron. The second neuron can stimulate the third neuron only if the former's axon is not inhibited by the first neuron.

Figure 2.7 is a schematic representation of a set of neurons. Each large circle represents the main body of a neuron, also known as the *soma*. The long extension from each soma represents an *axon*, which can convey a voltage signal down its length. At the end of the axon is a *terminal button*, shown in the figure as a darkened disk. The terminal button contains chemicals called neurotransmitters which can be released to stimulate the next neuron. The short extensions on each soma represent *dendrites*, or branches for receiving signals. Located on the dendrites are detectors called *spines*, represented in the figure as open disks.

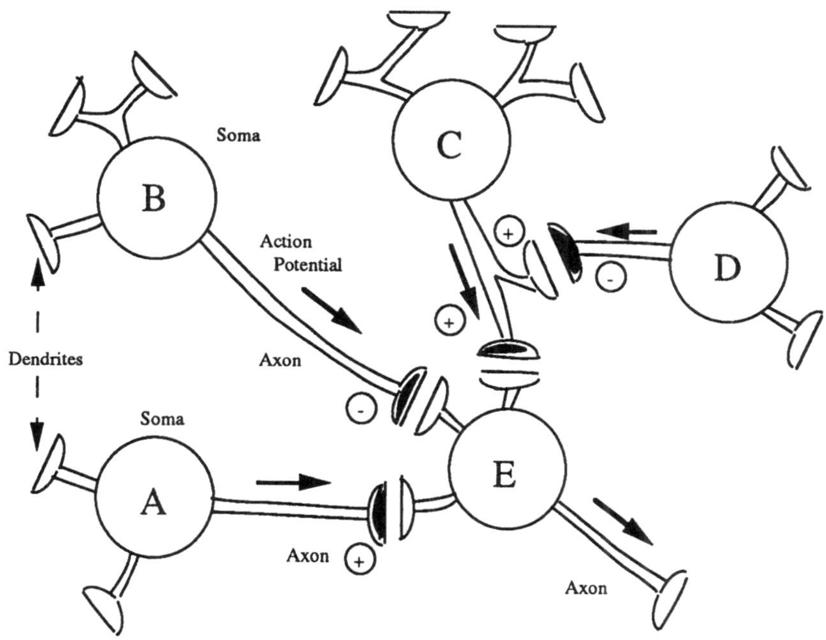

Figure 2.7 A network of neurons. Neuron A can send an excitatory signal to neuron E, while neuron B can issue an inhibitory signal. Neuron C can also excite neuron E, but only when the axon of neuron C is not itself inhibited by neuron D.

The gap between a terminal button and a spine is called a *synapse* or *cleft*. When a terminal button is activated, it releases neurotransmitters which diffuse across the synapse and attach to receptors on the button at the other end.

In the figure, neuron A can send an excitatory signal to neuron E. Neuron B, on the other hand, can inhibit neuron E. Neuron C can excite neuron E, but only if its axon is not inhibited by neuron D. In this way, the output of neuron D acts as a *gate* or switch for the output of neuron C.

The synapse between neurons A and E is called *axodendritic*, since the axon of A impinges on the axon of E. On the other hand, the axon of neuron D innervates the axon of cell C; hence the synapse is of the *axoaxonic* type.

Let the output o_j of neuron j equal 1 if its axon is active, and 0 otherwise. If the effect of neuron j on neuron i is excitatory, let the corresponding weight w_{ij} equal 1; but if the effect is inhibitory, let w_{ij} equal -1.

Let v_{jk} be the weight corresponding to an axoaxonic synapse from neuron k to neuron j. Then v_{jk} equals 1 if the synapse is excitatory, and -1 if inhibitory.

Let G_j^+ be the set of neurons having excitatory axoaxonic synapses impinging on neuron j. Similarly, let G_j^- be the set of cells having inhibitory axoaxonic switches.

Suppose that the output of axon j is active precisely when each excitatory switch is active, and each inhibitory switch is inactive. In other words, $o_k = 1$ if cell $u_k \in G_j^+$, and $o_k = 0$ if $u_k \in G_j^-$. Then the output of cell u_j is given by

$$o_j = a_j \prod_{k \in G_j^+} o_k \prod_{k \in G_j^-} (1 - o_k)$$

Let G_j denote the union of G_j^+ and G_j^-. By using the axoaxonic weights v_{jk}, we can also express the output of cell u_j in the form

$$o_j = a_j \prod_{k \in G_j} \left(\frac{1 + v_{jk} o_k}{2}\right)$$

The input to cell u_i is given by

$$v_i = \Sigma_j \, w_{ij} \, o_j = \Sigma_j \, w_{ij} \, a_j \prod_{k \in G_j} \left(\frac{1 + v_{jk} o_k}{2}\right)$$

The input to the cell is a weighted summation over dendrodendritic synapses, taking into account the product of outputs at the respective axoaxonic synapses.

The preceding formula involves both a summation and a product. Hence the underlying net is also called a *sigma-pi* configuration.

Weight Training

Part of the knowledge in a neural net is embedded in the activation function $f(a,v)$ and the output function o(a). A significant component, however, lies in the pattern of interconnections among the cells and their respective weights.

A suitable pattern of links and weights can be established through three activities:

1. Generation of new connections between pairs of cells.

2. Elimination of connections.

3. Adjustment of weights on existing connections.

The first two activities can be regarded as special cases of the third activity. More specifically, the generation of a new connection between cells i and j corresponds to the specification of a nonzero value for the weight w_{ij}, while the elimination of a link corresponds to the assignment of a zero value to w_{ij}.

Almost all learning rules are based on the notion of positive reinforcement: if the link between two cells is active, then its weight should be increased (Hebb, 1949). Let $w_{ij}(t)$ be the weight between cells i and j at time t. The value of the weight $w_{ij}(t+1)$ at the next step equals its previous value plus some increment $\Delta w_{ij}(t)$. We can write this relationship as

$$w_{ij}(t+1) = w_{ij}(t) + \Delta w_{ij}(t)$$

The salient quantity here is the increment $\Delta w_{ij}(t)$, which may be positive, negative, or zero. The increment may well depend on the extant weight $w_{ij}(t)$, as well as the output level $o_j(t)$ from the source node and the activation level $a_i(t)$ at the destination node.

We wish to allow for the possibility of exogenous input to the increment function $\Delta w_{ij}(t)$. This takes the form of a parameter $\tau_i(t)$, called the *teaching* input to the destination node. The level of the teaching input $\tau_i(t)$ can serve as a reference point for the weight w_{ij}; that is, the increment $\Delta w_{ij}(t)$ may be set higher or lower depending on the relationship of w_{ij} to τ_i.

Thus far we have identified the increment to be a function of the output $o_j(t)$, the weight $w_{ij}(t)$, the activation $a_i(t)$, and the trainer $\tau_i(t)$. In algebraic form, we have

$$\Delta w_{ij}(t) = \Delta w_{ij}\,(o_j(t),\, w_{ij}(t),\, a_i(t),\, \tau_i(t))$$

We can be more specific about the form of the increment function. In many cases, the function Δw_{ij} can be expressed as the product of two simpler functions: one depending on o_j and w_{ij}, and the other on a_i and τ_i. We can express this relationship as the product of two functions *r* and *s*:

$$\Delta w_{ij}(t) = r(o_j(t),\, w_{ij}(t))\; s(a_i(t),\, \tau_i(t)) \qquad \text{(Eq. 2)}$$

This equation can in turn be tailored to describe particular training algorithms. A number of such algorithms are described below.

Simple Hebbian rule. This is perhaps the most elementary type of learning rule. No training function τ_i is found in this algorithm, and the increment function ignores the extant weight w_{ij}.

From Eq. 2 we can specify the functions *r* and *s* as follows:

$$r(o_j(t), w_{ij}(t)) \equiv \eta\, o_j(t)$$
$$s(a_i(t), \tau_i(t)) \equiv a_i(t)$$

The constant parameter η is a scaling factor which represents the rate of learning. If η is large, the learning is rapid; otherwise the learning is gradual. (A large value of η may appear to be desirable at first sight. However, such large values may give rise to a suboptimal net, or even an unstable one.) The resulting increment function is

$$\Delta w_{ij}(t) = \eta\, o_j(t)\, a_i(t)$$

Teacher rules. This type of rule involves the use of a teacher. For one algorithm in this category, the functions *r* and *s* in Eq. 2 are given by

$$r(o_j(t), w_{ij}(t)) \equiv \eta\, o_j(t)$$
$$s(a_i(t), \tau_i(t)) \equiv \tau_i(t) - a_i(t)$$

The resulting increment function is

$$\Delta w_{ij}(t) = \eta\, o_j(t)\, [\tau_i(t) - a_i(t)]$$

This algorithm is called the *Widrow-Hoff* rule (Sutton and Barto, 1981).

Star rules. The term "star" derives from the stellar configuration of the pictorial representation, as shown in Figure 2.8 and 2.9 (Grossberg, 1974, 1976).

In the *instar* configuration, the increment Δw_{ij} for a link is proportional to the difference between the output level of the source node and the weight. We can set the factors in Eq. 2 as

$$r(o_j(t), w_{ij}(t)) \equiv o_j(t) - w_{ij}(t)$$
$$s(a_i(t), \tau_i(t)) \equiv \eta$$

The result is

$$\Delta w_{ij}(t) = \eta\, [o_j(t) - w_{ij}(t)] \qquad \text{(Eq. 3)}$$

The scaling factor η is often initialized to a value of 0.1 and gradually reduced toward zero during the training process.

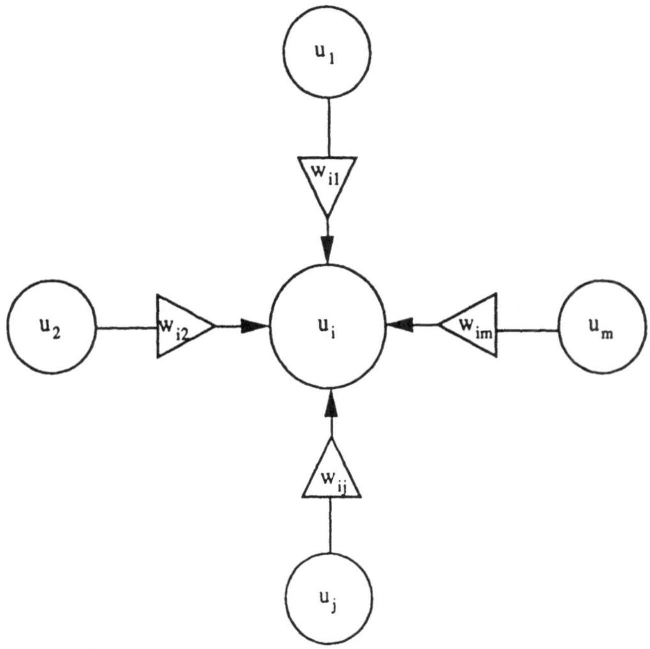

Figure 2.8 Instar configuration.

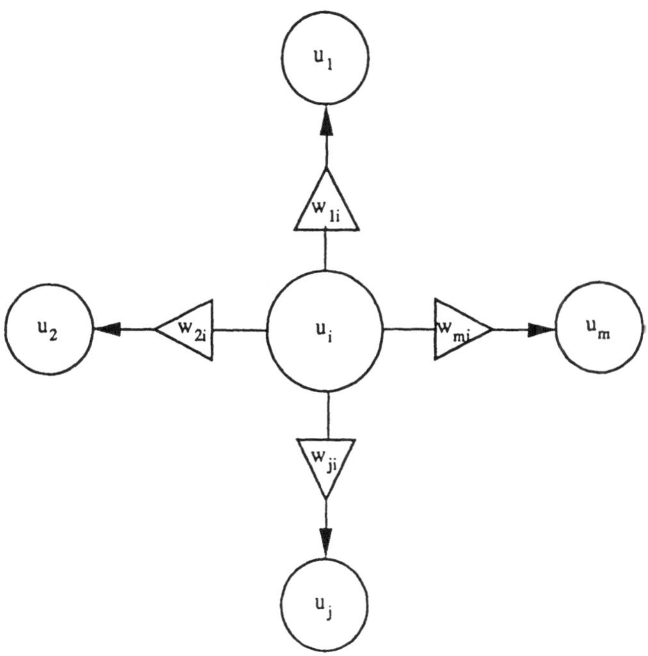

Figure 2.9 Outstar configuration.

Suppose the activation function simply equals the net input. In other words,

$$a_i(t+1) = \nu_i(t) \equiv \Sigma_j \, w_{ij}(t) \, o_j(t) \qquad \text{(Eq. 4)}$$

Then once the weights are properly set, the instar node u_i performs the standard function of pattern recognition. In algebraic terms, the activation $a_i(t+1)$ is maximal for a specific value of the input vector o(t).

The reason for this outcome is as follows. Suppose that the weight $w_{ij}(t)$ at time t is less than the output $o_j(t)$ from node j, which also happens to be the input to node i. From Eq. 3, we see that the increment $\Delta w_{ij}(t)$ will then be positive. Consequently, the new weight $w_{ij}(t+1)$ will be larger than before. The converse process will occur if the weight $w_{ij}(t)$ exceeds that of the output $o_j(t)$. In short, the output level $o_j(t)$ serves as the reference point toward which the weight w_{ij} will converge.

Let $w_i(t) \equiv (w_{i1}(t), w_{i2}(t),...,w_{im}(t))$ denote the vector of input weights to node i. From Eq. 4, the activation can be written as

$$a_i(t+1) = w_i(t) \, o(t)$$

where o(t) is the vector of input signals toward node i. In other words, the activation $a_i(t+1)$ is the inner or dot product of the vectors $w_i(t)$ and o(t). In geometric terms, this dot product is maximal when the two vectors are collinear. In short, the instar node will respond most strongly to the input vector o(t) which matches the weight vector $w_i(t)$.

We have noted that the training rate η is usually set to a low value such as 0.1, and decreased even further as the training progresses. If η were set to 1, we can deduce from Eq. 3 that $\Delta w_{ij}(t) = o_j(t) - w_{ij}(t)$ and the new weight will be

$$w_{ij}(t+1) = w_{ij}(t) + \Delta w_{ij}(t) = o_j(t)$$

In other words, the training will occur in a single step.

The disadvantage of such accelerated learning is a loss of the capacity to generalize. A gradual learning process results in the ability of the instar to recognize not only a particular input vector, but closely related ones as well. This property springs from the fact that multiple inputs can expose the net to the usual variations in the input vector. The weights are then determined by average values of the components in the input vector. Consequently, the net can easily recognize not only the "ideal" vector but routine variants of it as well.

The *outstar* configuration, shown in Figure 2.9, is the converse of the instar. Here the relevant links disperse from the central node u_i. In contrast to the instar which recognizes inputs, the outstar disseminates information to other nodes.

The training formula for the outstar is given by

$$\Delta w_{ij}(t) = \eta \, [o_i(t) - w_{ij}(t)]$$

When the training stabilizes, each outgoing weight w_{ij} will match a particular output level o_i. In this way, the outstar can impose its desired value on the outgoing links.

Training Procedures

In the previous section, we examined rules for updating the weights during a training process. This is, in some sense, the heart of the training procedure, but not the entirety. In particular, we must specify the initialization process as well as the termination condition.

Training algorithms may be classified into two types in terms of determinacy. In the *deterministic* category, the weights are changed in predictable fashion; in the *stochastic* category, the weights are modified in part by introducing probabilistic elements. We will consider examples of both types of training procedures.

Perceptron algorithm. The *perceptron* is a neural net consisting of two layers, the input and output levels, as shown in Figure 2.10. The structure has a *feedforward* organization: the signals move only forward, from the first to the second layer (Rosenblatt, 1959). A 2-layer net is severely limited in its range of abilities (Minsky and Papert, 1969). However, this type of net is of historical significance for having generated intense debate for and against neural nets.

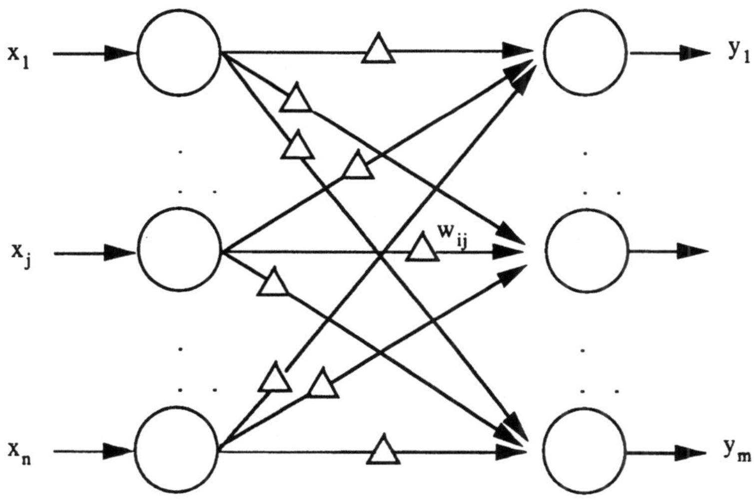

Figure 2.10 Perceptron network.

The perceptron relies on a closed-form learning procedure. Consequently, each input vector in the training set is matched with an output vector.

The components of the input vector are real numbers, while the output is a member of the binary set {0,1}. Intuitively, we can regard the role of the perceptron as taking a description of an object in the form of an input vector and determining whether it falls into one category or another.

The training procedure is as follows.

1. *Initialization*. Set each weight w_{ij} to a small random number.

2. *Input*. Present an input vector $x = (x_1, x_2,..., x_n)^T$ to the nodes in the input layer. Calculate the net input to the output layer as

$$v_i = \Sigma_j w_{ij} x_j$$

3. *Activation*. Compute the activation value a_i for each output node a_i from the unit step function $u(\theta_i, v_i)$ with parameter θ_i. Set the output value o_i to the activation value a_i.

4. *Error*. Calculate the error for each node. This is the difference between the target output $\tau_i \equiv y_i$, and the computed output o_i:

$$\varepsilon_i = \tau_i - o_i$$

5. *Weight update*. Modify each weight according to the following rule:

$$w_{ij}(t+1) = w_{ij}(t) + \eta x_j \varepsilon_i$$

6. *Termination check*. If the error is sufficiently small, stop; otherwise go to Step 2.

In this algorithm, the output level o_i is identical to the activation level a_i. Further, the input x_j equals the output o_j in the general notational framework. The error ε_i will be a member of the set {−1, 0, 1}, since τ_i and o_i are both binary numbers.

Widrow-Hoff algorithm. The perceptron was initially developed to handle only binary outputs; but this limitation was soon transcended. The improved approach uses a sigmoidal function to allow for continuous outputs. Another advantage of this method is a proof that the net will converge for any function that it can represent (Widrow, 1959; Widrow and Hoff, 1960).

The first model proposed by Widrow, called Adeline, exhibited a single cell in the output layer. A subsequent model, the Madeline, allowed for multiple cells.

The training procedure for the Adeline parallels that for the perceptron. The key difference lies in Step 4. Here the continuous variable v_i replaces the binary variable o_i. In other words, the error is calculated as

$$\varepsilon_i = \tau_i - v_i$$

where v_i is the net input to node i.

Deterministic, Closed-form Training

An example of a deterministic, supervised training algorithm is *backpropagation.* The goal of the algorithm is to produce a set of weights to ensure that the actual output *y* in response to an input *x* is close to some desired response *z.*

For the backpropagation procedure, the cells are arranged in a series of layers as shown in Figure 2.11. The initial layer is the *input layer,* which accepts the input vector $x = (x_1, ..., x_m)$. The input layer serves to distribute the incoming values x_i to the appropriate cells in the next layer; it performs no computation and therefore has no weights associated with it.

The last level in the network, tagged as layer M, is called the *output layer.* This layer contains n cells, corresponding to the output vector $y = (y_1,...,y_n)$. Under perfect circumstances, the output y would precisely match the desired target vector $z = (z_1,...,z_n)$ after the training is completed.

All other layers in the network are called *hidden layers*, since they do not interface directly with the external environment. In the figure, the generic weight w_{ik} lies on the arc between two adjacent hidden layers.

A convention in connectionist literature is to associate a weight w_{ji} with its terminal node u_j rather than its source node u_i. More loosely, we can associate the weight with the layer which contains the terminal cell u_j. For instance, the weight w_{ji} shown in Figure 2.11 is associated with layer M rather than layer p.

The backpropagation algorithm requires a differentiable set of transformation functions for each cell. Let the activation function for unit u_i be given by $a_i = f_i(v_i)$ where v is the net input to a cell. In line with previous notation, we will denote the output function as $o_i = g_i(a_i)$. We can compose these two functions by writing $o_i = g_i(f_i(v_i))$. The backpropagation procedure relies, in part, on the slope of the output with respect to the input:

$$\frac{d\,o_i}{d\,v_i} \equiv o_i' = \frac{d\,g_i}{d\,f_i} \bullet \frac{d\,f_i}{d\,v_i}$$

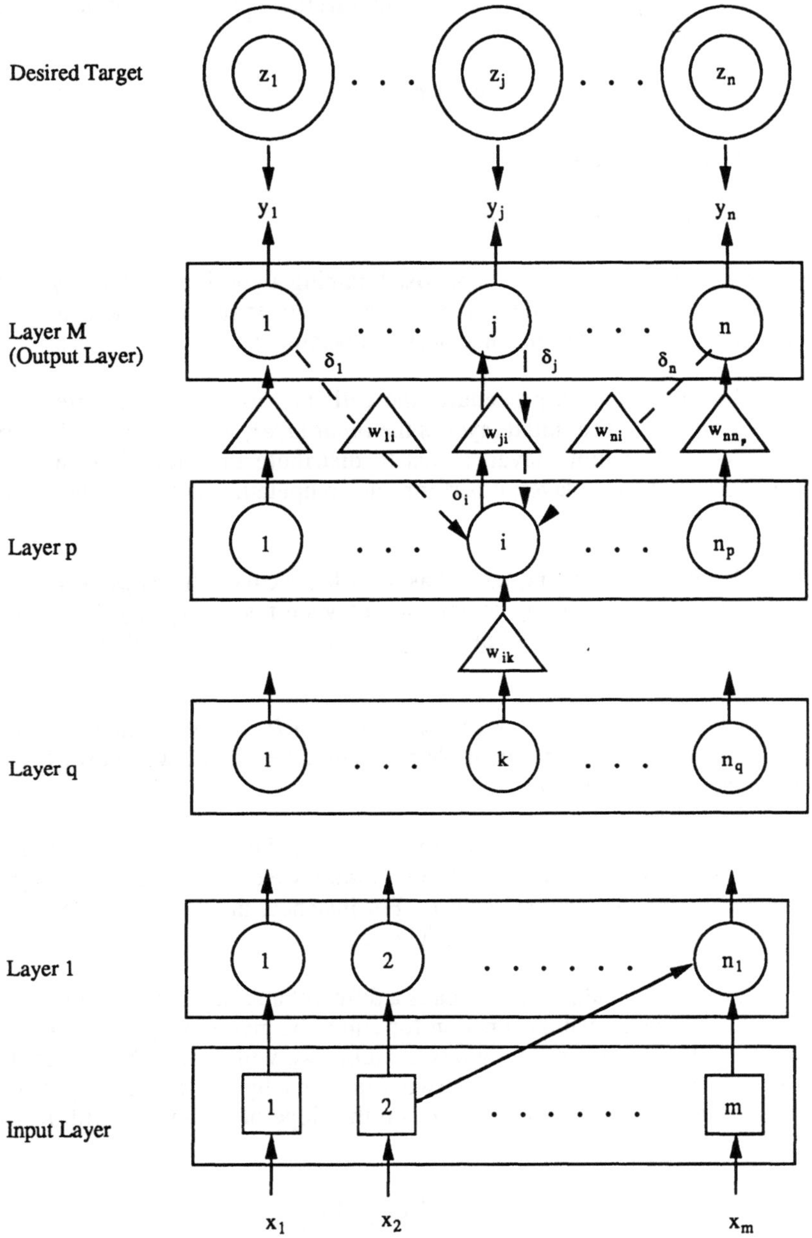

Figure 2.11 Layered configurations for backpropagation algorithm. Layers p, q and M are contiguous. For simplicity, only a few weights are shown explicitly.

To illustrate, let g_i be the identity function; and f_i the Boltzmann curve given by

$$f_i(v_i) = \frac{1}{1 + e^{-v_i}}$$

Then the slope of the output against input will be

$$o_i' = -(1 + e^{-v_i})^{-2} (-e^{-v_i}) = o_i (1 - o_i)$$

This relationship is often used in the training procedure, as explained below.

We finally turn to the backpropagation algorithm itself.

0. *Initialization.* Assign a small random number to each weight w_{ij}.

1. *Input.* Select a pair of input and target vectors; that is, an input vector x and the corresponding desired output z from the set of training data.

2. *Forward pass.*

   a. Compute the actual output y of the network based on the input x.

   b. Compute the error ε_j for each output node:

$$\varepsilon_j = z_j - y_j$$

   This is the discrepancy between the desired and actual output values.

3. *Backward pass.* Let $o_j(v_j)$ be the output function for cell u_j based on the input v_i.

   a. *Output layer.* Compute the adjusted error or discrepancy δ_j. This results from weighting the error ε_j by the slope o_j' of the output against the input. In symbolic form:

$$\delta_j \equiv o_j'(v_j)\varepsilon_j \qquad \text{(Eq. 5)}$$

   For instance, suppose that the output function is given by the Boltzmann relation: $o_j(v_j) = (1 + \exp(-v_j))^{-1}$. From a previous discussion, we can deduce that $\delta_j = o_j(1 - o_j)\, \varepsilon_j$. The change in weight w_{ji} is proportional to the discrepancy δ_j at the terminal cell and the output value o_i from the source cell (see Figure 2.11). More specifically, the increment Δw_{ij} is given by

$$\Delta w_{ij}(t) = \eta \, \varepsilon_j \, o_i \tag{Eq. 6}$$

The training rate η indicates the speed at which changes are effected; a common range for this parameter is 0.01 to 1.0. This computation is performed for each weight associated with the output layer.

b. *Hidden layer(s).* Each hidden layer is trained by using the error associated with the subsequent layer. The error at the output layer is propagated back through the entire network, training the hidden layers one by one.

The discrepancy δ_j for each cell in the output layer was calculated using Eq. 5. In the backward pass stage of the training process, this information is propagated through the network in the reverse direction.

At this juncture, the weights w_{ji} associated with the output layer M have been incremented. The discrepancy δ_j at layer M is considered as the output of the source cells in the reverse propagation. The incoming information to cell u_i in layer p is indicated by the dotted arrows in the figure. In algebraic terms, the net input μ_i to cell i in the reverse process is

$$\mu_i = \sum_j w_{ji} \, \delta_j$$

The discrepancy δ_i for cell u_i is given by the product of the reverse net input and the slope o_i' of the output function. In other words,

$$\begin{aligned} \delta_i &= o_i' \, \mu_i \\ &= o_i' \sum_j w_{ji} \, \delta_j \end{aligned} \tag{Eq. 7}$$

Here δ_i belongs to the hidden layer and δ_j to the subsequent layer, as shown in Figure 2.11.

Once the discrepancy δ_i is computed for each cell in the hidden layer, its associated weight w_{ik} can be incremented using

$$\Delta w_{ik} = \eta \, \delta_i \, o_k \tag{Eq. 8}$$

The next layer is trained by treating the current hidden layer as if it were the output layer. The calculations are similar: Eq. 7 is used to compute the discrepancy for each cell, and Eq. 8 to determine the associated weights. The reverse pass continues until the first layer has been trained.

4. *Termination.* If the error between the actual and desired outputs is sufficiently small, stop the training. Otherwise go to Step 1.

The initialization process at Step 0 calls for the assignment of small, random numbers to each weight. The smallness ensures that the network avoids saturation by powerful signals. The randomness avoids certain other pathologies; for instance, if the initial weights are all equal, then the training could fail when the desired output involves unequal values.

Accelerated learning. The backpropagation algorithm can be remarkably slow in yielding acceptable levels of performance. One technique to reduce training time and enhance stability is to take into account the weight change at the preceding step. More specifically, the *momentum* method involves the following relation (Rumelhart, Hinton, and Williams, 1986):

$$\Delta w_{ik}(t) = \eta\, \delta_i\, o_k + \alpha\, \Delta w_{ik}(t-1)$$

The first term is simply that due to Eq. 8. The second term involves the weight increment $\Delta w_{ik}(t-1)$ at the previous step. This quantity is scaled by the momentum factor α, often set to a value of roughly 0.9. In other words, the momentum coefficient assigns an inertia to the velocity of weight changes. The utility of the momentum method depends on the application; it improves learning behavior in some cases but hinders in others.

Another technique is that of *exponential smoothing* (Sejnowski and Rosenberg, 1987). This involves a weighted combination of the two operational factors in the preceding equation:

$$\Delta w_{ik}(t) = \eta\, \left[(1-\alpha)\delta_i\, o_k + \alpha\, \Delta w_{ik}(t-1)\right]$$

The smoothing factor α takes a value in the interval [0,1]. At the lower extreme of $\alpha=0$, the increment $\Delta w_{ik}(t)$ depends entirely on the product of the discrepancy δ_i and output o_k, moderated by the training rate η. In contrast, the upper limit of $\alpha=1$ implies that the new increment depends only upon the old increment, attenuated by the rate η.

Open-form Training

Most training algorithms are of the closed-form variety, in which an input vector x is paired with a desired output vector z. In open-form training, however, the system examines a set $\{x^i\}$ of data and attempts to extract any inherent regularities. One such methodology relates to self-organizing maps used for pattern recognition activities.

A self-organizing map classifies a pattern in the form of an input vector. Each component of the vector represents a single aspect of the underlying domain, be it a segment of a photograph or the condition of mechanical equipment. When the training is completed, an input vector from a given category will produce an identifiable output in each cell in the output layer. The neuron with the greatest excitation represents the specific category.

As mentioned previously, the training is conducted without a target vector. Consequently there is no way to determine *a priori* which output cell will align itself with a particular category. But this can be deduced readily by testing the net after the training has terminated.

The training procedure regards the n possible weights on input arcs to each cell i as a vector in n-dimensional space. In other words, $w_i \equiv (w_{i1}, w_{i2}, \ldots, w_{in})$ represents the vector of incoming weights on unit i.

To initialize the training process, each component of the weight vector is assigned a random number. Then the weight vector is normalized to ensure unit length. More specifically, let $v_i = (v_{i1}, v_{i2}, \ldots, v_{in})$ denote the initial vector of random weights for unit i. Then the normalized weight vector is

$$w_i = \frac{1}{\|v_i\|} v_i$$

where $\| v_i \|$ denotes the Euclidean norm of v_i.

Every input vector x' is normalized to unit length according to the same procedure; that is, $x \equiv x' / \| x \|$.

The training procedure is as follows (Kohonen, 1984):

1. *Input.* Apply an input vector $x = (x_1, x_2, \ldots, x_n)$.

2. *Discrepancy.* Compute the distance δ_i between x and the weight vector w_i of each cell i. The Euclidean distance is given by

$$\delta_i = \| w_i - x \| = \left[\sum_{j=1}^{n} (w_{ij} - x_j)\right]^{\frac{1}{2}}$$

3. *Selection.* The cell whose weight lies closest to x is regarded as the winner. The corresponding weight vector w* serves as the center of a collection of weight vectors w_k within a δ-neighborhood. In other words, each w_k lies within a distance δ of the central weight w*.

4. *Adjustment.* Each weight w_k in the δ-neighborhood is moved toward x by a fraction α. The algebraic transformation is

$$w_k(t+1) = w_k(t) + \alpha\, (x - w_k(t))$$

5. *Termination*. If all input vectors x have been exhausted, stop; otherwise return to Step 1.

During the training process, both the δ-neighborhoods and the adjustment factors are reduced. This is accomplished by reducing the values of δ and α. The initial value of δ can be as large as the maximal distance between weight vectors, then gradually diminish so as to include only one other weight vector w_k; a reasonable value for α is close to 1 at the outset and a gradual decline toward 0.1. For effective results, the number of training cycles should be at least 500 times the number of output cells (Kohonen, 1984).

The geometric interpretation for this procedure is as follows. Since all the vectors are normalized to unit length, they may be viewed as points on a unit sphere in n-dimensional space. The selection stage in Step 3 identifies the weight w* which lies closest to the input vector x. Among the weights depicted in Figure 2.12, vectors w_1 and w_4 lie within a δ-neighborhood of the central weight w^*.

The adjustment process in Step 4 shifts the neighboring weights toward the vector x. The difference vector $(x - w_1)$ is the one reaching from the head of w_1 to the head of x. An adjustment factor of $\alpha = 0.5$, for instance, would mean that w_1 is moved halfway toward x after the transformation is implemented. A decrease in the factor α corresponds to an attenuation of the adjustments in the weights w_k as the training progresses. A fall in the value of δ implies that the "sphere of influence" of the central vector w^* diminishes over time. In this way, the adjustments tend to stabilize during the course of the training.

The training process is predicated on the assumption that the input vectors cluster naturally into neighboring regions. In Figure 2.12, the vectors p and q might both represent a single category, and x a second category. A particular class will tend to dominate a specific neuron; the corresponding weight will be shifted toward the center of that category. As a result, an input vector from that category will excite the respective neuron to the greatest extent.

Once training is completed, an arbitrary input vector is applied to the net. The excitation level for each neuron is computed, and the highest level identified. The corresponding cell is taken to be the indicator of the correct category for the input vector.

Probabilistic Algorithms

In this subsection, we examine training methods based on nondeterministic techniques. These procedures are also called *probabilistic* or *statistical* methods.

Probabilistic techniques make use of random changes in the values of weights. In general, the changes which result in improved performance are retained while other perturbations are ignored. A short-term degradation in performance might lead to a system configuration which ultimately accommodates long-term effectiveness; consequently, a change which results in degraded performance is occasionally retained with nonzero probability.

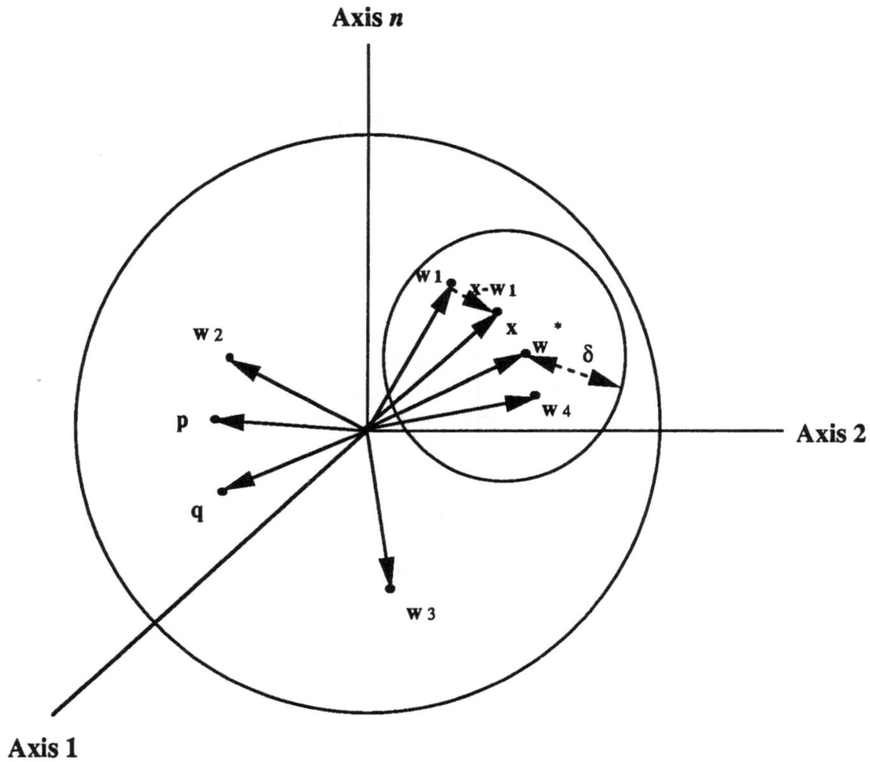

Figure 2.12 Weights and input vectors as points on a unit sphere in n-dimensional space.

Let $x^i = (x_1{}^i, x_2{}^i, ..., x_n{}^i)$ be an input vector and $z^i = (z_1{}^i, z_2{}^i, ..., z_n{}^i)$ the desired output vector. Assume that $y^i = (y_1{}^i, y_2{}^i, ..., y_n{}^i)$ is the actual output produced by the network. A closed-form training procedure attempts to modify the weights in a network in such a way that the actual output y^i in response to the input x^i is sufficiently close to the desired vector z^i.

Various metrics can be used to indicate the discrepancy between the actual and desired outputs. One such measure, called the *Manhattan distance*, simply scans the discrepancies in all dimensions for any pair of vectors y and z:

$$\varepsilon_1 \equiv \sum_{i=1}^{n} | y_i - z_i |$$

A more common metric is the squared error, given by

$$\varepsilon_2 \equiv \sum_{i=1}^{n} (y_i - z_i)^2 = (y - z)^T (y - z)$$

In geometric terms, ε_2 represents the square of the distance between vectors y and z in n-dimensional space.

The error metric is also called the *performance measure* or *objective function*. The goal of the training procedure is to improve system performance until the error metric falls below some threshold θ_ε; that is, until $\varepsilon < \theta_\varepsilon$ for whatever metric ε is being utilized.

A probabilistic algorithm for closed-form training takes the following form:

1. *Output.* Apply an input vector x^i to the network; calculate the output vector y^i based on the current configuration of weights.

2. *Error.* Compute the discrepancy between the actual output y^i and the desired vector z^i, using some error metric ε.

3. *Perturbation.* Select an arc at random and change its weight by a small amount. Compute the new output y^i corresponding to the same input x^i as well as the corresponding error $\tilde{\varepsilon}$.

   a. If the new error $\tilde{\varepsilon}$ is smaller than the old error ε, then retain the change.

   b. If the new error $\tilde{\varepsilon}$ is not smaller, then retain the change anyway with a probability obtained from some probability distribution function. Otherwise, return the weight to its previous setting.

4. *Termination.* If the error is sufficiently small, stop; otherwise go back to Step 1.

This procedure tends to minimize the error metric. As with most extremization techniques, however, the procedure can result in entrapment at a local rather than a global minimum.

Figure 2.13 illustrates the problem for a one-dimensional situation; namely, one involving only a single weight w. In the figure, a local minimum for the error metric ε occurs at w_1, while a global minimum exists at w_2. Suppose that the current value of the weight is w_0. An attempt to decrease the error metric ε will cause the weight to be perturbed toward w_1, much like a ball rolling downhill. Once w_1 is reached, however, any small perturbation in w will be associated with a rise in the error metric ε. Hence the training algorithm will settle at the local minimum lying at w_1 rather than the global minimum at w_2. This is the problem with small perturbations in the weight w.

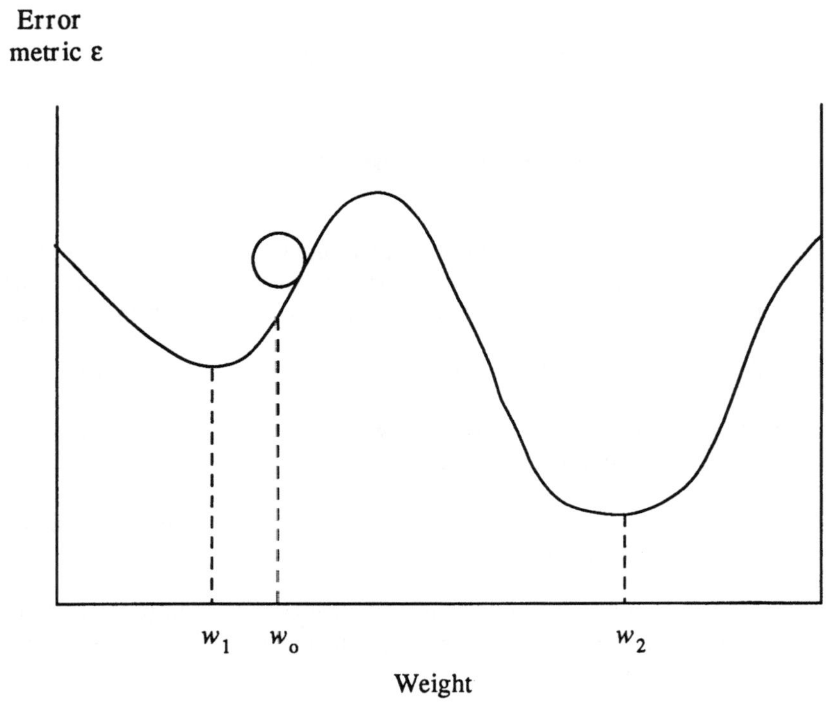

Figure 2.13 Example of a nonlinear function having a local minimum at w_1 and a global minimum at w_2.

On the other hand, suppose that large perturbations are employed. Then the weight will visit the vicinity of both w_1 and w_2 with great frequency; but every other value of w is as likely to be visited. Consequently, the weight will jump erratically from one value to another without settling down to the desired minimum at w_2.

To circumvent this dilemma, the technique of *attenuated perturbations* is used; we begin with large perturbations in the weight, then gradually decrease the magnitude of the changes. In this way, the system can stabilize after avoiding local optima.

A physical interpretation for attenuated perturbations is as follows. Consider the circle in Figure 2.13 to represent a marble in a box, lying on a surface defined by the sinuous curve; the curve may be viewed as a two-dimensional surface extending into the page. Suppose the box is shaken vigorously from side to side along the plane of the page. Then at any instant, the marble is equally likely to lie over any point on the surface.

If the violence of the shaking is reduced, there will come a time when the marble has enough energy to escape from the shallow valley at w_1 into the deep valley at w_2, but not vice versa. At this juncture, the marble is caught in the neighborhood of the global optimum at w_2. As the vigor of the shaking gradually vanishes, the marble will eventually settle at the global minimum at location w_2.

The method of attenuated perturbations has a physical analogy in the annealing of metals, and is therefore also known as *simulated annealing*. Consider a metal whose temperature exceeds the melting point. As with any physical system, the atoms in the metal will tend toward a configuration of minimum energy. The high temperature, however, keeps the atoms moving violently and prevents the attainment of this energy minimum. As the metal cools and the temperature gradually falls, the atoms attain configurations of decreasing energy. Eventually the atoms can form a single crystal whose energy is the lowest of all the available configurations. (The salient aspects of statistical physics are discussed further in the Appendix.)

The distribution of energy levels in the annealing process takes the following form:

$$p(E) \propto \exp[\, -E / (kT)\,]$$

Here p(E) is the probability density function as a function of the energy E; k is Boltzmann's constant; and T is the absolute temperature. When the temperature T is very large, the right hand side of the expression approaches 1 for all energy states; that is, a high energy state is about as likely as a low energy state. However, when the value of T is small, the term $e^{-E/kT}$ falls very rapidly with increases in the energy E; in other words, the probability of high energy states is small, and the system is most likely to find itself in a low energy configuration.

A training algorithm based on simulated annealing is called the Boltzmann procedure.

Boltzmann procedure. This algorithm consists of the following steps:

1. *Initialization.* Define a parameter T which reflects some kind of artificial temperature. Initialize T at a high value.

2. *Output.* Apply an input x^i to the network, and calculate the output y^i as well as the error ε.

3. *Perturbation.* Select an arc at random; change its weight randomly; calculate the new output $\tilde{y}^i$ and the corresponding error $\tilde{\varepsilon}$.

4. *Termination.* If $\tilde{\varepsilon} < \varepsilon$, retain the weight change. Otherwise, retain the change with a probability from a suitable exponential distribution function. If $\tilde{\varepsilon}$ is sufficiently small, stop; otherwise decrease T and go to Step 2.

The Boltzmann distribution function is employed at the perturbation step above. If the new error $\tilde{\varepsilon}$ is smaller than the old error ε, then the perturbation is retained. Otherwise, even if $\tilde{\varepsilon} \geq \varepsilon$, the perturbation is still retained with positive probability. Let $\delta \equiv \tilde{\varepsilon} - \varepsilon \geq 0$ be the increase in error. The probability of accepting the perturbation is

$$p(\delta) = \frac{1}{kT}\, e^{-\delta/kT}$$

Here T is the artificial temperature, while k is an analogue of Boltzmann's constant. The value of k must be selected for the application at hand, but the value of T is gradually reduced in the process of simulated annealing.

The Boltzmann procedure requires that the weight changes in Steps 3 and 4 be considered for *each* link. The random perturbation in Step 3 can be effected in various ways. In analogy with thermal systems, one possibility is the Gaussian distribution:

$$p(\Delta w) = \frac{1}{T\sqrt{2\pi}} \exp\left(-\frac{\Delta w^2}{2T^2}\right)$$

where Δw is the perturbation.

The temperature parameter T must be decreased at a particular rate for effective results. More specifically, the reduction in temperature must be proportional to the reciprocal of the logarithm of time:

$$T(t) = \frac{T_0}{\log(1+t)}$$

where t is the artificial time, and T_0 the initial temperature. This will ensure the attainment of a global minimum in the error function (Geman and Geman, 1984).

The low rate of cooling is a liability of the standard Boltzmann procedure. In practice, the slow convergence often results in untenable computational requirements.

To circumvent this problem, the Cauchy distribution can be substituted for the Gaussian distribution. This accelerates the Boltzmann procedure dramatically, allowing for a linear cooling rate (Szu and Hartley, 1987):

$$T(t) = \frac{T_0}{1+t}$$

The rapid convergence results from the fact that the Cauchy distribution is symmetric like the Gaussian, but has thicker tails as shown in Figure 2.14. Consequently, the Cauchy distribution permits larger perturbations more frequently than the Gaussian rule.

Specific Heat Method. In addition to the use of the Cauchy distribution, another way to accelerate the cooling process springs from the notion of phase changes in physical systems.

The *specific heat* of a thermal system is defined as the rate of temperature change due to an energy change. The annealing of a metal results in phase changes. The specific heat often changes abruptly during a phase transition, as the system settles into a local energy minimum.

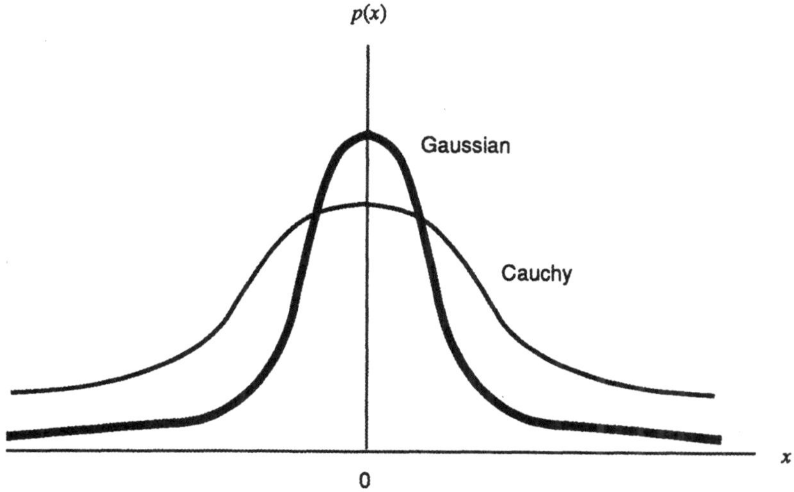

Figure 2.14 Comparison of Gaussian and Cauchy distributions.

Artificial nets are subject to similar phase changes and the associated discontinuities in specific heat. The *artificial specific heat* can be defined as the average rate of temperature change with respect to the error metric.

Consider again the marble-in-a-box example. The perturbations are vigorous at the outset. Consequently, the average value of the error metric is roughly independent of small variations in temperature; therefore the specific heat undergoes little change. At the opposite extreme of low temperatures, the system is locked into a minimal energy configuration, so the specific heat is roughly constant. In these extreme regions, the temperature can be varied rapidly, since the error function is largely unaffected.

At critical junctures, however, a small temperature change can yield a dramatic shift in the error metric. In Figure 2.13, the temperature where the marble is likely to go from w_1 to w_2, but not the converse, is a critical point; here, the error metric can change abruptly. At a critical point such as this, the temperature must be reduced slowly to ensure that the marble avoids entrapment at location w_1.

The critical values of temperature can be monitored as abrupt changes in the specific heat. At such points, the temperature must be reduced gradually. In other regions, the cooling can be accelerated.

In the next section, some key features of neural nets are illustrated in the context of a pattern associator.

Pattern Associator

The function of a neural net is epitomized in a *pattern associator*, which assigns a set of outputs y with a collection of inputs x. To illustrate, the inputs might be the set of signed 4-tuples; in symbolic form $x = \{(x_1, x_2, x_3, x_4) \mid x_i = -1 \text{ or } 1\}$. In conjunction, the set of outputs might be binary pairs; that is, $y = \{(y_1, y_2) \mid y_i = -1 \text{ or } 1\}$. The function of the pattern associator is to map x onto y.

In particular, consider the input vector $x = (1, -1, 1, 1)^T$. Suppose that this input is to be associated with the output vector $y = (1, -1)^T$. What set of weights will result in this transformation?

Since x has four degrees of freedom and y only two, there are many solutions. Two such solutions are as follows:

$$W' = \begin{bmatrix} 1 & 0 & 0 & 0 \\ 0 & -1 & 0 & 0 \end{bmatrix}$$

$$W = \begin{bmatrix} 1/4 & -1/4 & 1/4 & 1/4 \\ -1/4 & 1/4 & -1/4 & -1/4 \end{bmatrix}$$

For instance, the required relation for the second solution can be verified by the matrix equation

$$\begin{bmatrix} 1 \\ -1 \end{bmatrix} = \begin{bmatrix} 1/4 & -1/4 & 1/4 & 1/4 \\ -1/4 & 1/4 & -1/4 & -1/4 \end{bmatrix} \begin{bmatrix} 1 \\ -1 \\ 1 \\ 1 \end{bmatrix}$$

A pictorial representation is given in Figure 2.15.

Graceful Degradation

A neural network fails softly rather than catastrophically under degradation. The *graceful degradation* refers both to noisy input as well as internal damage.

To illustrate, consider again the neural network of Fig. 2.15. Suppose that the first term of the input were changed by noise from a + 1 to a −1. In other words, the degraded input is $x' = (-1, 1, 1, 1)^T$. The new output is then $y' = (1/2, -1/2)^T$; this differs from the original output of $y = (1, -1)^T$, but the character of the output is still recognizable in the positive and negative signs.

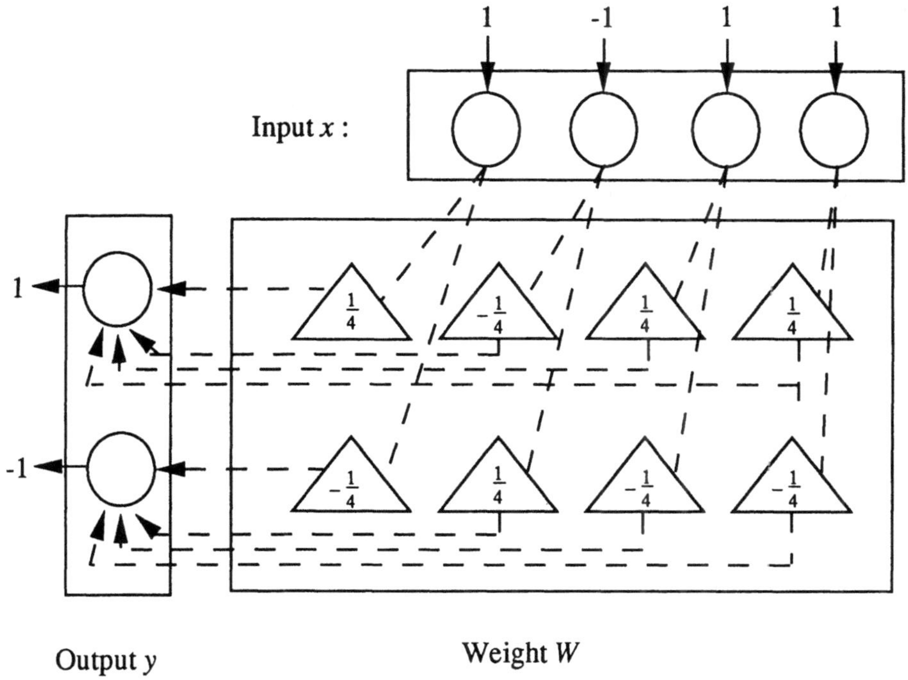

Figure 2.15 Example of a pattern associator, matching $x=(1,-1,1,1)^T$ to $y=(1,-1)^T$. A degradation in the input, such as $x'=(-1,-1,1,1)^T$, or minor damage to the weights, will cause only limited error in the output y.

Graceful degradation under system damage can be illustrated as follows. Suppose that the very first weight, w_{11} were destroyed, thereby preventing the corresponding signal from reaching the y_1 node (in algebraic terms, this is equivalent to setting $w_{11} = 0$). Assume that the original input $x = (1, -1, 1, 1)^T$ is used. Then the new value of output will be $y'' = (3/4, -1)$, rather than the original value of $y = (1, -1)$. If the output is known to take only binary values, then the error can be easily rectified and the proper input and output identified.

Suppose that both the input and system were degraded. For the preceding examples, we have $x' = (-1, -1, 1, 1)^T$ and $w_{11} = 0$. The resulting output will be $y''' = (3/4, -1/2)$, which again exhibits the proper nature of the output.

The property of graceful degradation in a neural network springs from the distributed nature of knowledge. Since the knowledge is spread across many components of the system, a localized failure will result in only a marginal change from the original behavior.

Biological Systems

Biological systems display a richness of structure and function. This versatility is reflected in various systems, including the visual apparatus.

Vision System

Much of the learning for an animal occurs through its visual system. In an immediate context, the visual system allows the animal to learn about its present environment, including the location of food, shelter, and predators. Over the longer term, vision allows for observation and learning about the behavior of objects in the environment, both animate and inanimate.

In higher animals, the visual system shows a high degree of organization. The basic unit of processing is the neuron, but neurons take many structural forms and perform diverse functions.

Anatomy of Vision

Figure 2.16 shows visual association areas on the left hemisphere of the rhesus monkey brain (Carlson, 1991; Kandel and Schwartz, 1985; Zeki, 1978). The association areas on the right hemisphere are similar.

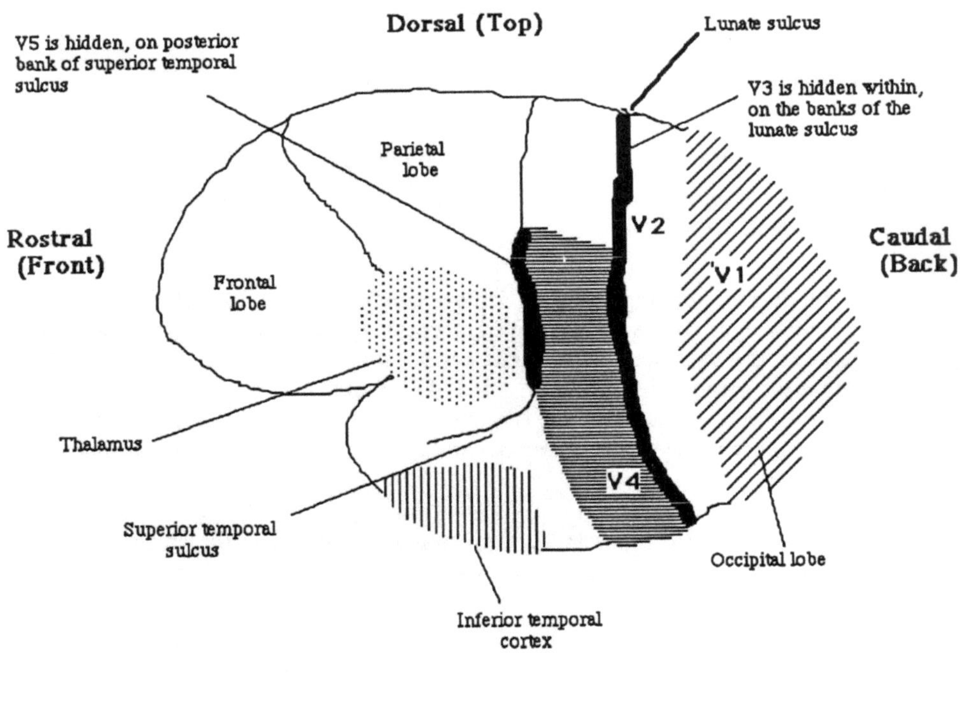

Figure 2.16 Schematic of the visual association cortex for the rhesus monkey brain.

Optic nerves extend from the retina of each eye to the thalamus, a relay center located deep within the brain. The thalamus consists of two lobes, one in each hemisphere of the brain. More specifically, the optic nerves enter the part of the thalamus known as the dorsal lateral geniculate nucleus. Here the nerves synapse with a second set of neurons – the optic radiations – which send signals to the primary visual cortex.

The primary visual cortex, also called area V1, performs elementary signal processing. The processed information is then sent to other areas of the occipital lobe, namely V2, V3, V4, and V5.

Areas V3 and V5 are not shown explicitly in Figure 2.16. Between regions V2 and V4, for instance, is a fissure, or sulcus, created by a folding of the cerebral cortex. Area V3 lies on the cerebral cortex at both sides of the sulcus, thereby being obscured from direct view. A similar situation holds for area V5, which lies on the posterior surface of the superior temporal sulcus.

After processing in the occipital lobe, visual information moves to other areas for further processing. Recognition of complex objects, such as hands or faces, occurs in the inferior temporal cortex. In contrast, spatial information, such as the relative location of a pair of objects, is recognized in the parietal lobe.

Architecture of Vision

The architecture of the visual system is depicted in Figure 2.17 (Carlson, 1991; Livingstone and Hubel, 1988). The ganglion cells which extend from the retina to the dorsal lateral geniculate nucleus are of two major types: magnocellular and parvocellular.

The magnocellular subsystem, found in all mammals, is evolutionarily older. It responds quickly to visual stimuli and is sensitive to low levels of contrast. However, it is color-blind and has low spatial resolution. In contrast, the parvocellular system has high spatial resolution and is color-sensitive; but it is slow and has poor sensitivity to contrast. The information corresponding to each subsystem maintains its separate identity as it travels to the primary visual cortex.

Elementary visual processing occurs in area V1. For instance, a type of neural unit called a *simple cell* detects lines of a particular orientation and location in the visual field. Another type of detector, called a *complex cell*, responds best to the movement of a line in a particular direction.

The information from area V1 is conveyed to V2 for further processing. For example, some neurons in V2 respond well to differences in retinal images and thereby appear to form the basis for binocular vision.

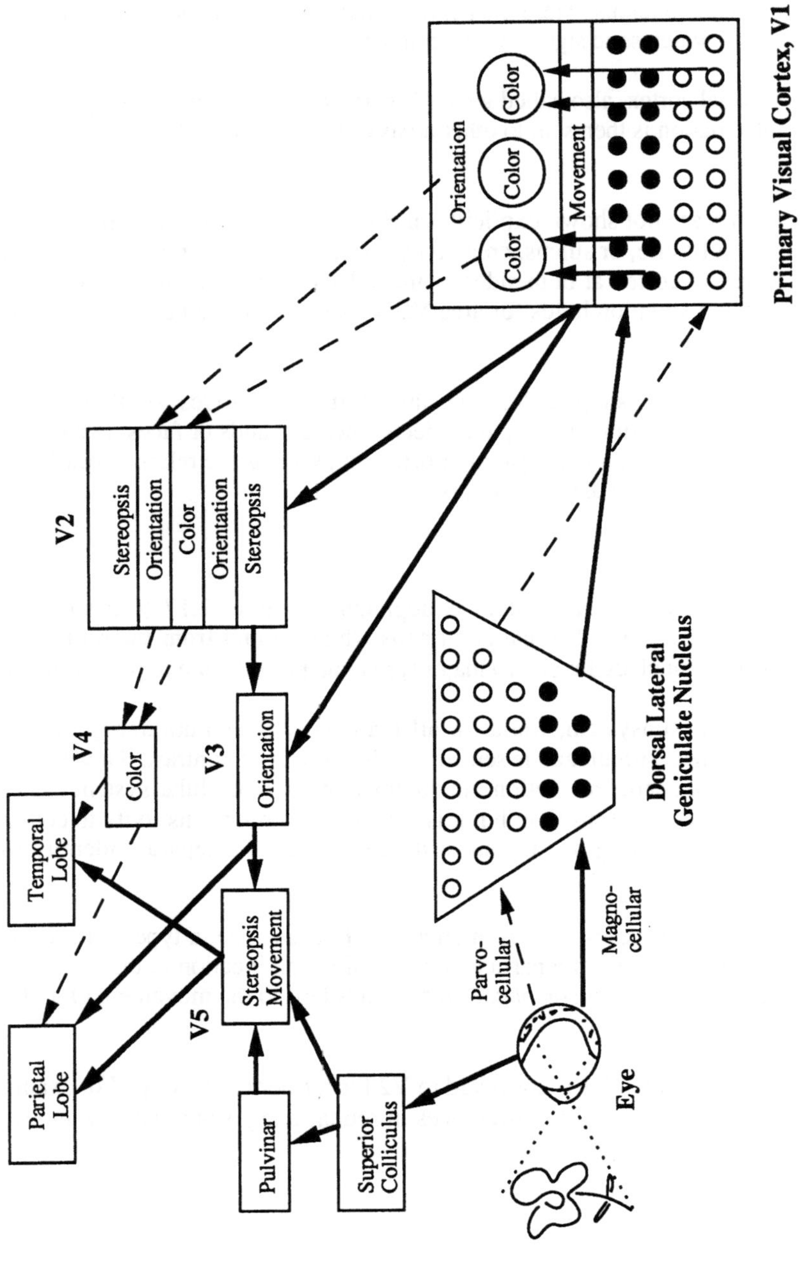

Figure 2.17 Architecture of the visual system for primates.

Area V3 is responsible for integrating orientation information; it additionally serves as a precursor to V5, which handles the analysis of movement. Area V5 also receives information from the superior colliculus, a protrusion in the brain near the thalamus. Information from the superior colliculus also travels to the pulvinar, a component of the thalamus, then on to area V5.

Area V4 is responsible for the interpretation of color. In conjunction with area V5, region V4 projects neurons into the parietal and temporal lobes. As we have seen before, the parietal lobe is responsible for spatial reasoning and the temporal lobe for high-level recognition of complicated features.

Implications for Artificial Systems

The visual apparatus epitomizes the functional decomposition of biological systems. The architecture of the visual system illustrates a hierarchy of information processing, from low to high levels.

Further, functional partitioning occurs even within a particular level. For instance, the neural circuitry within the optic nervous system is retinotopic; that is, information from adjacent points in the visual field fall on neighboring ganglion cells in the retina, and this correspondence is maintained in transit through the dorsal lateral geniculate nucleus and beyond.

Not only is there a functional partitioning of activities within the brain, but a variety of hardware components. This is exemplified by the diversity of neural types within the retina. Although the preceding discussion has largely ignored the issue of the physical characteristics of neurons, the visual processing activities rely on numerous types of cells.

If artificial nets are to approach the speed and versatility of biological systems, it seems imperative to introduce hybrid structures within the system. The hybrid organization relates to functional decomposition as well as hardware elements specialized for particular tasks.

Not only learning speed, but the ultimate capabilities of the mind may well spring from the hybrid structure of the brain:

> The marvelous powers of the brain emerge not from any single, uniformly structural connectionist network but from highly evolved arrangements of smaller, specialized networks which are connected in very specific ways.
> (Minsky and Papert, 1988, p. xv.)

We have examined the visual system of primates as an example of the multi-level structure of biological networks.

A Hybrid Organization for Artificial Networks

Long term memory appears to be encoded in microstructural changes in neurons. In this sense, the database is flat, consisting of only one physical level: namely, that of neurons.

On the other hand, memory at the knowledge level has both associative and hierarchical aspects. It is associative in that the word "red" may evoke images of blood, roses, *The Scarlet Letter*, valor, love, or any number of other objects and concepts. But knowledge is also hierarchically organized; for instance, an organism can be viewed as a collection of organ systems, each of which consists of organs, in turn comprised of tissues, then cells, then organelles, and so on. The hierarchical organization manifests itself even in information processing at the cortical level: a set of pixels might be recognized as lines, then classified as an angle which forms part of a pentagon, etc.

What is the implication for neural nets? To date, much of the work in connectionism has regarded neural nets as a single flat layer, consisting of multiple copies of identical processors. Other approaches have used a sequentially staged approach, in which information from one layer of cells is sent on to a layer containing cells of another type.

In contrast, it may be advantageous to group these processors into hierarchical as well as associative clusters. The benefits lie in efficiency and robustness, as well as explicitness of knowledge representation.

The *efficiency* would result from localized processing. An item of knowledge need not be dispersed over the entire network, much as in a holographic recording, but specialized within a module. One advantage is that the processors can learn to develop localized procedures; as an example, axons are myelinated and have large diameters when speed of transmission is crucial. In addition, localized processing obviates the need for long-distance communication among processing units as a matter of routine.

The argument for localized processing should not be taken as a counterclaim to some holographic characteristics of biological networks. The brain is holographic in the sense that knowledge of complex images or procedures appears to be distributed over multiple neurons *within a module* and occasionally across a few modules. This arrangment allows for properties such as graceful degradation and associative recall from partial information. The distributed nature of knowledge at the level of a module comes as no surprise, since a single neuron is likely to be limited in the amount of information it can encode.

The hierarchical decomposition may also be crucial in supporting *explicitness* of knowledge. A liability of connectionism is that knowledge is implicit. We have no way of looking at a set of neurons and deducing that "It snowed yesterday" is encoded in the referent individual; and the same is true for synthetic networks. As the information theorist would say, information is lost in the transformation from a high-level to a low-level representation. If the hierarchical organization is sufficiently fine grained, however, such deduction may be possible. For instance, one type of linkage from the "yesterday" component of the "temporal relationships" module to the "snow" component of the "precipitation" module might denote the relationship in question.

A third advantage of multi-level hierarchical decomposition lies in *robustness*. By closely clustering related concepts, it becomes feasible to explore additional ideas merely by examining the local neighborhood. This efficient examination of related concepts can help prevent the problem of fragility and overspecialization. In other words, a global rather than local optimum is more likely to be found. Moreover, if the system encounters a new environment which strains its capabilities, it can examine its knowledge base systematically by reaching out radially in the hierarchy, both up and down layers and within a single layer. Consequently, the most relevant knowledge will likely be tapped first, followed by concepts of decreasing utility.

Efficiency in Weights

Let N_T denote the total number of computations required during the training period. This number depends both on the number L of parameters or links as well as the number C of computations per link. More specifically, N_T is the product of the two other variables:

$$N_T = LC$$

One way to reduce N_T is to reduce the number C of computations per weight. Another way to save on computational resources is to reduce the number L of links. This can be achieved by a modular architecture.

A hierarchical structure can result in a dramatic decrease in the required number of weights or linkages. As an example, consider a net consisting of $n = 64$ units. A fully connected network allowing for bidirectional arcs will require $n^2 = 4096$ weights. Suppose instead that the units are clustered successively into quadruples, as indicated in Figure 2.18. In other words, four units are grouped into a single module. Next, four of these first-level modules are grouped into a second-level module.

Suppose that each module is fully connected, as a single entity, to every other module at the same level. What is the required number of links? There are four units within each first-level module. These would require $4^2 = 16$ links to be fully connected. Since we have $n = 64$ units in all, there are 16 modules at level 1. The total number of links among elementary units is then $L_0 = 16(4^2) = 256$.

In a similar way, there are 4 first-level modules within each second-level module. This would imply $4^2 = 16$ linkages. Given 4 second level modules, the total number of links at level 1 is $L_1 = 4(4^2) = 64$. Finally, there are 4 modules at level 2, requiring $L_2 = 4^2 = 16$ links for full connectivity.

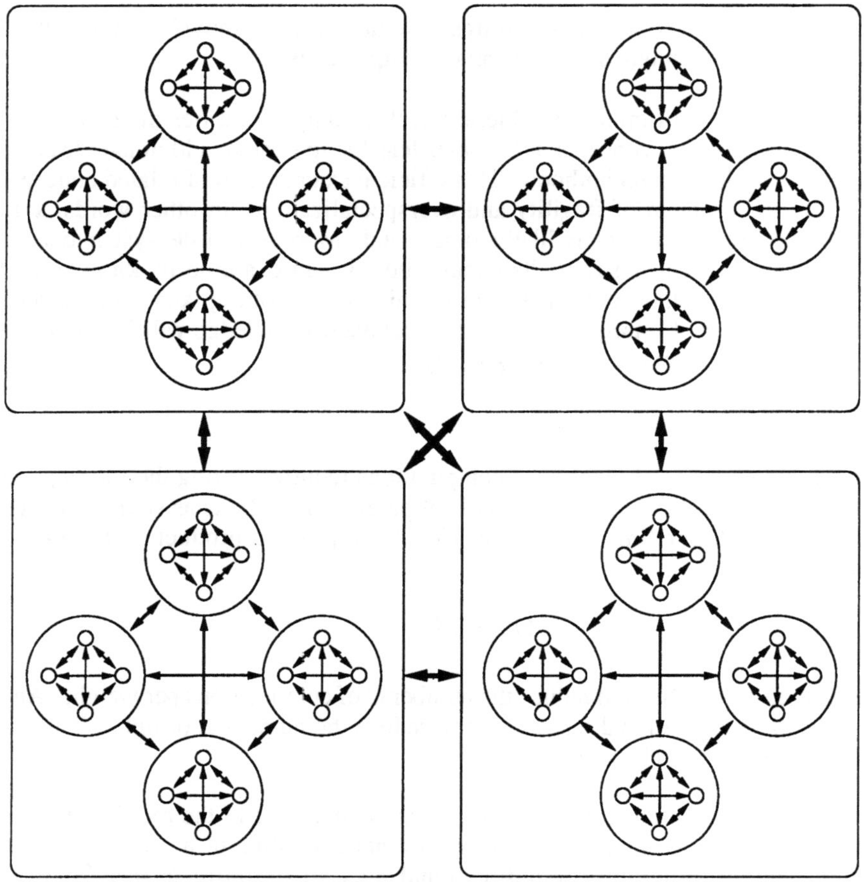

Figure 2.18 Example of a hierarchical organization for processing units. The required number of links and weights is an order of magnitude smaller than for a "degenerate" hierarchy consisting of a single level.

In all, we have

$$L = L_0 + L_1 + L_2 = 336$$

links. This number is smaller by more than an order of magnitude than the 4096 links needed to connect the original structure. In terms of the required number of links and associated weights, a hierarchical structure can yield dramatic improvements over the "flat" organization consisting only of a single level.

Effect on Computational Speed

As suggested in an earlier section, reducing the number of links will save on computational resources. The throughput or turnaround time for a net, however, is determined by the number of layers rather than simply on the number of cells. Hence reducing the number L of links will not necessarily reduce the turnaround time between input and output. Rather, the enhanced speed for a hierarchical organization springs primarily from the ability to draw on high-level descriptors or features.

To illustrate, the design of an automobile is much simplified by employing high-level descriptors such as a specific engine from a standard set, or a particular wheel from a standard stock. This approach obviates the need to reason about the car in every minute detail such as the dimensions, tolerances, finishes, and so on for each part. In fact, reasoning at such low levels of detail would quickly overload the mental capacities of a human designer, and result in computational intractability for a computer of reasonable capacity.

Biological systems, in constrast, allow for a hierarchy of feature representations. For instance, certain cells in the auditory cortex respond only to positive phase or temporal differences in signals reaching the two ears; and others only to negative differences. As another instance, we have seen that movement in a particular direction is detected by complex cells in area V5 of the primate visual cortex; but sophisticated features such as faces are recognized in the parietal cortex.

Such a hierarchy of features allows for flexibility in reasoning about the world. Moreover, the ability to draw on high-level features eases the computational burden. In this way, a hierarchical organization of neural nets is more likely to approach the speed of biological systems. In fact, since the speed of electronic transitions in a silicon chip is on the order of nanoseconds while neural impulses require milliseconds, ample opportunity exists for a suitably sophisticated artificial network to exceed the speed of its biological counterpart.

Efficiency in Operation

Once a neural net is trained, it can operate with refreshing swiftness. The speed follows from the fact that the number of computational stages required is determined by the number of layers in the net, rather than the number of processing elements. For instance, if a neural network has 3000 units but only 3 layers, the computations can be performed in only three stages.

The speed of operation has been used to advantage in a number of implementations. One class of applications relates to the simulation of natural systems in real time for constrained domains. An example involves the role of the hippocampus in temporal and spatial navigation. A neural network was designed to simulate the computations of pyramidal cells of the hippocampus in forecasting movement as a function of time and space. The network was effective in emulating the trajectories followed by real mice in reaching for food (Schmajuk, 1990).

A hierarchical communication structure may or may not correspond to a hierarchical processing scheme. To illustrate, consider two separate modules which process information from visual and auditory sensors. Each module may process its own information while exchanging intermediate results with the other module. The input from one module serves to constrain the space of potential solutions generated by the other. In fact, the mutual interaction of constraints might be the only way to disambiguate among candidate solutions. For instance, a moving figure in the shadows might be a dog or a cat; it emits a noise like the wail of a baby or a cat; hence it must be a cat. In this case, the processing of visual and auditory signals can occur in parallel, whether or not each sensory mode is relegated to different modules of the neural system.

Conclusion

A comparison of natural and artificial neural networks highlights structural as well as operational differences. Many types of algorithms have been proposed for artificial networks, yet they remain slow and clumsy compared to natural processes.

A key differentiator seems to lie in the structural sophistication of biological systems. Moreover, depending on their respective functions, organic cells are often richly interconnected and arranged as well into specialized modules.

These observations suggest avenues for improving the performance of synthetic systems. By building on the diversity of structures and processing modes of biological mechanisms, engineered systems hold the promise of matching – and eventually surpassing – their natural counterparts.

References

Anderson, Dana Z. *Neural Information Processing Systems*. New York: American Institute of Physics, 1988.

Anderson, J.A. and Rosenfeld, E. (eds) *Neurocomputing*. Cambridge, MA: MIT Press, 1988.

Arbib, Michael A. and Hanson, Allen R., eds. *Vision, Brain, and Cooperative Computation*. Cambridge, MA: MIT Press, 1987.

Borg-Grahm, Lyle J. "Simulations Suggest Information Processing Roles for the Diverse Currents in Hippocampal Neurons." In Anderson (1988), pp.82-94.

Carlson, Neil R. *Physiology of Behavior*, 4th ed. Boston: Allyn and Bacon, 1991, ch. 6.

Gefenstette, J.J. *Proc. International Conference on Genetic Algorithms and their Applications*. Carnegie-Mellon University, Pittsburgh, PA, July 1985.

Geman, Stuart and Donald Geman. "Stochastic Relaxation, Gibbs Distributions, and the Bayesian Restoration of Images." *IEEE Trans. on Pattern Analysis and Machine Intelligence*, v. 6(6), 1984, pp. 721-741.

Goldberg, D.E. Computer-aided gas pipeline operation using geneticalgorithms and rule learning. Ph.D. thesis, University of Michigan, 1983.

Grossberg, S. "Adaptive Pattern Classification and Universal Recoding: Part I. Parallel Development and Coding of Neural Feature Detectors." *Biological Cybernetics*, v. 23(3), 1976, pp. 121-134.

Grossberg, S. "Classical and Instrumental Learning by Neural Networks." *Progress in Theoretical Biology*, v. 3. NY: Academic Press, 1974, pp. 51-141.

Haken, Herman, ed. *Neural and Synergetic Computers.* New York: Springer-Verlag, 1988.

Hebb, D. O. *The Organization of Behavior.* NY: Wiley, 1949.

Holland, J.H. *Adaptation in Natural and Artificial Systems.* Ann Arbor, MIUniv. Michigan Press, 1975

Holland, J.H. Escaping brittleness: the possibilities of general-purpose learning algorithms applied to parallel rule-based systems. In R.S.Michalski et al., eds., *Machine Learning: An Artificial Intelligence Approach, v. II*, Los Altos, CA: M. Kaufmann, 1986: 593-624.

Holland, J.H. and J.S. Reitman. Cognitive systems based on adaptive algorithms. In D.A. Waterman and F. Hayes-Roth, eds., *Pattern-Directed Inference Systems*, NY: Academic, 1978: pp. 313-329.

Hopfield, J. J. "Neural Networks and Physical Systems with Emergent Collective Computational Abilities." *Proc. Nat. Acad. Sciences USA*, v. 79(8), 1982, pp. 2554-2558.

Livingstone, M. and D. Hubel. "Segregation of Form, Color, Movement and Depth: Anatomy, Physiology, and Perception." *Science*, v. 240, 1988, pp. 740-749.

Kandel, Eric R. and James H. Schwartz. *Principles of Neural Science.* 2nd ed. NY: Elsevier, 1985, pp. 344-395.

Kim, S.H. A mathematical framework for intelligent manufacturing systems. *Proc. of Symposium on Integrated and Intelligent Manufacturing Systems*, ASME, Anaheim, CA, Dec., 1986.

Kim, S.H., "A Unified Framework for Self-Learning Systems". *Proc. Manufacturing International '88, v. III: Symp. on Manufacturing Systems,* Atlanta, GA, April 1988: 165-170.

Kim, S.H. *Designing Intelligence.* NY: Oxford University Press, 1990.

Kirkpatrick, S., C. D. Gelatt, Jr., and M. P. Vecchi. "Optimization by Simulated Annealing." *Science*, v. 220(4598), 1983, pp. 671-680.

Kohonen, T. *Self-Organization and Associative Memory.* NY: Springer-Verlag, 1984.

Kompass, E.J. and T.J. Williams, eds. Learning Systems and Pattern Recognition. Barrington, IL: Technical Pub., 1983.

Langley, P., G.L. Bradshaw, and H.A. Simon. Rediscovering chemistry with the BACON system. In R.S. Michalski, J.G. Carbonell and T.M. Mitchell, *Machine Learning: An Artificial Intelligence Approach*, Palo Alto, CA: Tioga, 1983: 307-329.

Lenat, D. EURISKO: a program that learns new heuristics and design concepts: the nature of heuristics, III: program design and results. Artificial Intelligence, v.21(2), 1983: 61-98.

Mangel, Marc. "Evolutionary Optimization and Neural Network Models of Behavior." Journal of Mathematical Biology, v.28 (3), 1990, pp.237-256.

Minsky, Marvin and Seymour Papert. *Perceptions: An Introduction to Computational Geometry*. Cambridge, MA: M.I.T. Press, 1969; expanded ed., 1988.

Nadel, Lynn, Cooper, Lynn A., Culicover, Peter, and Harnish, R. Michael, eds. *Neural Connections, Mental Computation*. Cambridge, MA: MIT Press, 1989.

Pao, Y-H. "A Connectionist Net Approach to Autonomous Machine Learning of E Effective Process Control Strategies." *Robotics and Computer-Integrated Manufacturing*, v.4 (3/4)1988, pp. 633-642.

Porter, Bruce W. and Mooney, Ray J., eds. Machine Learning: Proceedings of the Seventh International Conference on Machine Learning. Palo Alto, CA: Morgan Kaufmann, 1990.

Rosenblatt, F. *Principles of Neurodynamics*. NY: Spartan, 1962.

Rumelhart, D.E., G.E. Hinton and J.L. McClelland. "A General Framework for Parallel Distributed Processing." In D.E. Rumelhart, J.L. McClelland and the PDP Research Group, *Parallel Distributed Processing*, v.1: Foundations, Cambridge, MA: M.I.T. Press, 1986, pp. 45-76.

Sammut, Claude and Cribb, James. "Is Learning Rate a Good Performance Criterion for Learning?" In Porter and Mooney (1990), pp.170-178.

Schmajuk, Nestor A. "Role of the Hippocampus in Temporal and Spatial Navigation: An Adaptive Neural Network." Behavioral Brain Research, v. 39, 1990, pp.205-229.

Shibazaki, H. and S.H. Kim. "Learning Systems for Manufacturing Automation: Integrating Explicit and Implicit Knowledge." *Robotics and Computer-Integrated Manufacturing*, v. 9, 1992.

Sutton, Richard S. and Andrew G. Barto. "Toward a Modern Theory of Adaptive Networks: Expectation and Prediction." *Psychological Review*, v. 88(2), 1981, pp. 135-170.

Szu, Harold and Ralph Hartley. "Fast Simulated Annealing." *Physics Letters A*, v. 122(3,4), 1987, pp. 157-162.

Tsypkin, Y.Z. Adaptation and Learning in Automatic Systems, trans. by Z.J. Nikolic.NY: Academic, 1971.

Wasserman, Philip D. *Neural Computing: Theory and Practice*. NY: Van Nostrand Reinhold, 1989.

Widrow, B. "Adaptive Sampled-Data Systems, A Statistical Theory of Adaptation." in *1959 IRE WESCON Convention Record*, part 4. NY: Institute of Radio Engineers, 1959.

Widrow, B. and M. Hoff. "Adaptive Switching Circuits." in *1960 IRE WESCON Convention Record*. NY: Institute of Radio Engineers, 1960.

Zeki, S.M. "The Cortical Projections of Foveal Striate Cortex in the Rhesus Monkey." *J. of Physiology*, v.277, 1978, pp. 227-244.

Principles of Coordination 3

The effectiveness of a system relates to its ability to fulfill its functional requirements. When a system consists of a set of collaborating agents, the overall performance may depend more on their ability to work together than on optimizing individual behavior. In fact, the goal of an agent is often at odds with the interest of the collective. The limitations of centralized control for complex systems suggest the need for decentralized coordination. This can be achieved by having each agent take explicit account of the cost of engaging in an activity as well as a measure of reward for competent behavior. The consideration of payoffs and penalties leads to an economic perspective of multiagent systems. In consequence, it is possible to draw on previous work in diverse fields, ranging from game theory to welfare economics and social policy. The promise and limitations of the explicit valuation approach are examined. To illustrate, the behavior of collaborative systems can be interpreted in terms of games of strategy; this approach permits a better understanding of the conditions for globally optimal behavior, as well as strategies for their attainment. The notions of agents and explicit valuation of action are versatile concepts. One indication of the versatility lies in the fact that these concepts cover as a special case the idea of genetic algorithms as a mechanism for learning systems.[1]

Introduction

The construction of systems of increasing complexity raises the challenge of, and underscores the need for, techniques for system design. As a way to contain complexity, many systems are engineered using a policy of modularization. Often systems of practical consequence, ranging from telecommunication networks to automated factories and even individual robots, consist of a set of interacting components. In a communication network, one node must relay its message packet to another if its usual neighbor is temporarily delayed with electronic traffic. In a factory, one production cell may be selected over another if its benefit-cost ratio is more favorable. Within an autonomous robot, the mobility system must work together with the manipulator component and the sensory apparatus if the machine is to deliver its payload properly.

[1] The presentation in this chapter is based on Kim (1991).

The components of a system may be subject to varying levels of autonomy. A collection of mobile robots may operate autonomously, for instance, or be subject to varying degrees of direction from a central controller.

In this chapter, we define an ***agent*** as any system or component capable of making decisions. This represents a very liberal view of autonomy. But it is consistent with the fact that few, if any, systems are entirely independent. Even humans in a democratic society are bound by laws, conventions and mores which constrain their space of behaviors.

We first examine the advantages and limitations of centralized control for coordinating multiple agents. To address these limitations, a decentralized approach based on individual initiative and economic criteria is described. More specifically, each agent takes explicit account of payoffs for its contribution toward collective objectives, and of the associated cost of performance.

Limitations of Centralized Control

Centralized control is a pervasive scheme for coordinating system behavior. It is reflected in the operating system of a conventional computer, the central nervous system of an animal, the coordination of a flexible manufacturing cell, the hierarchical structure of a company, and the authority of a federal judicial system.

Perhaps the greatest appeal of centralization lies in its simplicity. One authoritative center determines what needs to be done, and imposes its will on the rest of the system.

The centralized scheme, however, is subject to various limitations. The most prominent liabilities are the following (Kim, 1990):

1. *Complexity*. As the number of agents increases, the complexity of the system grows combinatorially. For a system of n agents, the number of potential pairwise interactions grows as $n(n-1)/2$. In general, however, interactions can involve any number of agents. The number of potential groups grows as 2^n, not to mention the variety of potential interactions among groups of groups.

2. *Timeliness*. As the tempo of the environment increases, the allowable response time for the system decreases. In fact, the tempo may be so high as to preclude the possibility of communication with the central node before undertaking a timely response. For instance, a drilling machine faced with a broken tool should terminate its operation immediately to prevent further damage to the workpiece or the machine itself; there is little time to await commands from a central controller.

3. *Vulnerability*. Control is vested in a single, central node. If this node is disabled, the entire system grinds to a halt.

4. *Inefficiency*. Only for a simple system can a central controller take full account of each agent's status and environmental circumstances, both spatially and temporally. In general, the agents will face different circumstances on one floor of a factory versus another, or on one continent versus another. Further, the environment as well as the status of individual agents will change over time. In a larger context, the inability to take advantage of – or even adjust to – local conditions may have been the single most important factor in the failure of centralized economic planning in the communist countries of Eastern Europe in the twentieth century.

Decentralization based on Explicit Valuation

The limitations of centralized control highlight the promise of decentralized mechanisms for coordinating complex systems. In fact, each type of liability associated with centralized control can be addressed, to a greater or lesser extent, through decentralization:

1. *Complexity*. An agent has to be concerned with global objectives and individual behavior. The agent might also need to keep track of some gross characteristics of other agents, such as the actions of other agents that have decided to interact with the former. But it need not be concerned with a wealth of other information, such as the complete status of other agents, their decision procedures, their environmental circumstances, or potential modes of interaction.

2. *Timeliness*. An agent can make decisions by itself or in interaction with associated agents, without the need to communicate with a single, overburdened, central node.

3. *Vulnerability*. Capability is distributed among the entire population of agents. The failure of any one agent will not necessarily jeopardize the mission of the system.

4. *Inefficiency*. Each agent can keep track of its own circumstances, and their evolution over time. These circumstances relate to internal status as well as environmental conditions.

In a decentralized approach, each agent has relative autonomy over its own actions. The agent is guided by a knowledge of overall system objectives as well as its own status and the consequences of its behavior.

In general, each agent will have only partial knowledge of the intentions and decisions of fellow agents. An agent's behavior is determined by its awareness of the payoffs and penalties due to each action. The use of a performance criterion in terms of net utility is tantamount to a financial perspective. In other words, each agent can be regarded as an economic entity that seeks to maximize its reward and minimize its cost. The reward is determined by the success to which an agent contributes to the system-level goals. The cost, on the other hand, arises from the agent's activities, and may be attributed to factors such as energy consumption, use of capital, time expenditure, and other resources depending on the application.

The explicit valuation of benefits and costs allows us to draw on previous models and results from fields ranging from game theory to welfare economics and social choice theory. The way in which this may be done is the subject of the subsequent sections.

Games and Strategies

A *game* is a situation where several agents interact (Aubin, 1979; Friedman, 1986; Ichiishi, 1983; Karlin, 1959; McKinsey, 1952; Rasmusen, 1989; Shubik, 1982; Vorob'ev, 1977). Let $N = \{1, 2, \ldots, n\}$ denote the set of agents or *players*, and A_i the set of actions or strategies available to player i. A strategy vector $a = (a_1, a_2, \ldots, a_n)$ is the list of actions due to each agent in the game.

Associated with any strategy vector a is a corresponding payoff or *performance measure* π_i to the ith player. In other words, $\pi_i = \pi_i(a)$. The goal of the agent is to maximize his payoff π_i.

Table 3.1 Pure conflict: zero-sum game.

		Americans	
		Land	Sea
British	Land	(–2,2)	(3,–3)
	Sea	(3,–3)	(–2,2)

For a game involving two players, the strategies and outcomes can be readily represented in tabular form. Table 3.1 represents a game of pure conflict. The British may attack by land or sea, while the Americans also have the same options for defense. The rows indicate the strategy set A_1 = {Land, Sea} available to the British, while the columns denote the strategy space A_2 = {Land, Sea} for the Americans. The entries in the table indicate the corresponding payoffs (π_1, π_2) where the subscript 1 refers to the British and 2 the Americans. For instance, suppose the British attack by sea while the Americans also lie in wait along this route; then the British will be defeated, yielding $\pi_1 = -2$ and $\pi_2 = 2$. On the other hand, if the British arrive over land and surprise the Americans, the latter will be defeated, resulting in $\pi_1 = 3$ and $\pi_2 = -3$.

This situation exemplifies a *zero-sum* game in which the payoffs in each cell add to zero. That is,

$$\pi_1(a_i, a_j) + \pi_2(a_i, a_j) = 0$$

for each strategy vector (a_i, a_j). In a zero-sum game, one player's gain is the other player's loss.

An example of pure cooperation is reflected in Table 3.2. Suppose that the Japanese and Europeans are contemplating a universal communications standard for consumer electronic products. If both players settle on the Alpha protocol, or both on Beta, then each party will reap a payoff of 9 units. On the other hand, if they select differing standards, sales of products will be limited and each player will gain only 1 unit of payoff. The players must obviously agree on a protocol for their mutual benefit. This is a clear case of a cooperative game.

Table 3.2 Pure cooperation.

		Europeans	
		Alpha	Beta
Japanese	Alpha	(9,9)	(1,1)
	Beta	(1,1)	(9,9)

We now consider a problem called the Prisoner's Dilemma. In this scenario, two suspects have been apprehended for a crime and separated from each other, thereby preventing communication. The choice for each suspect is as follows: if one confesses and his partner does, too, they both get 10 years in prison. However, if one confesses and testifies against the other, who does not, then the cooperator will go free while the other receives a sentence of 20 years. If they both refuse to cooperate with the authorities, they can only be convicted of minor crimes resulting in a sentence of 1 year. The appropriate payoffs are given in Table 3.3.

Table 3.3 Prisoner's Dilemma

		Suspect 2	
		Confess	Refuse
Suspect 1	Confess	(–10,–10)	(0,–20)
	Refuse	(–20,0)	(–1,–1)

Which action should be chosen? The consequences for each suspect are such that confession leads to a shorter sentence whether his partner confesses or not. By this reasoning, each suspect would choose to confess. The outcome of this strategy is a sentence of 10 years for each suspect.

In contrast, if both parties were to refrain from confessing, each would receive only 1 year. Here is a situation in which individual rationality is at odds with group rationality. The root of the dilemma lies in the lack of communication between the two players.

The strategy vector (Confess, Confess) is an example of a local optimum. This situation is called a ***Nash equilibrium,*** a strategy vector in which each player's action is the best for him given the action of his rival. As indicated above, a Nash equilibrium is not necessarily a global optimum.

An example of a game involving continuous action variables lies in the following scenario. Consider two firms producing identical products. Let q_1 and q_2 denote the respective quantities produced by each agent. The price of the good is a decreasing function of the overall quantity on the market:

$$p = 120 - q$$

where $q \equiv q_1 + q_2$ denotes the overall production. The corresponding chart is shown in Figure 3.1. We will assume that the marginal cost of production is negligible, as exemplified by the bottling of mineral water.

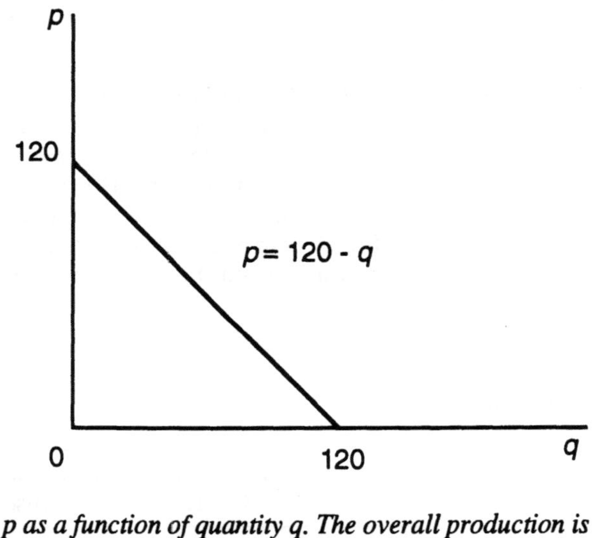

Figure 3.1 Price p as a function of quantity q. The overall production is $q = q_1 + q_2$, *where* q_i *is the production due to firm i. The payoff to firm i is* $J_i = pq_i$

Suppose that the quantity q_2 due to firm 2 is regarded as fixed. What is the first firm's best action? The profit to firm 1 is

$$\pi_1 = pq_1 = [120 - (q_1 + q_2)]\, q_1$$

Setting the first derivative to zero yields

$$\frac{\partial \pi_1}{\partial q_1} = 120 - 2q_1 - q_2 = 0$$

Solving the second equality for q_1 yields the best action for firm 1:

$$q_1 = \frac{120 - q_2}{2}$$

Since the second derivative $\partial^2 \pi_1 / \partial q_1^2$ is negative, this stationary point is a maximum. By symmetry, the best quantity for firm 2 is

$$q_2 = \frac{120 - q_1}{2}$$

The same symmetric argument implies that $q_1 = q_2$, from which we infer that $q_1 = q_2 = 40$. Consequently, the market price is $p = 40$. The payoff to each firm is therefore $\pi_i = pq_i = 1{,}600$. These results are tabulated in Table 3.4.

Table 3.4 Equilibrium strategies corresponding to Figure 3.1.

Eqilibrium	q_1	q_2	q	p	J_1	J_2
Nash	40	40	80	40	1,600	1,600
cooperative	30	30	60	60	1,800	1,800

This outcome is based on individual rationality. The production rate $q_i = 40$ is a Nash equilibrium representing the best action for each firm when the other's strategy is assumed to be fixed.

The Nash equilibrium, however, is not a global optimum. The overall payoff to the group is given by

$$\pi = pq = (120 - q)q$$

Setting the first derivative to zero yields

$$\frac{\partial \pi}{\partial q} = 120 - 2q = 0$$

from which $q = 60$. If the production rate were to be shared equally, the resulting quantities would be $q_1 = q_2 = 30$. The price would be $p = 60$, and each firm would earn $\pi_i = pq_i = 1{,}800$. This payoff exceeds the value of 1,600 resulting from the Nash equilibrium.

The suboptimality of the Nash solution springs from the myopic nature of the assumptions. Each firm seeks to maximize its own profit while regarding its rival's strategy as fixed, thereby ignoring the fact that the other firm will react in turn.

Here is another situation where individually rational behavior is collectively suboptimal. In multiagent systems, it is imperative to consider the consequences to the group as a whole.

Deadlock among independent systems

In general, the information available to each agent will differ from that accessible to follow decision makers. In a fully decentralized system devoid of a coordinating center, the agents may make suboptimal decisions. In some cases, the group may be unable to reach a decision at all.

As an illustration, consider a group of three agents named Al, Bo and Cy. The decision problem concerns the appropriate level of a parameter. The candidate actions are to increase, decrease, or maintain the current value of the parameter:

Action 1 : Increase
Action -1 : Decrease
Action 0 : Maintain

The parameter may relate to any number of situations: the level of temperature in a polymer production process, the length of time allowed for a psychological experiment, the inflation rate to maintain a healthy economy, and so on.

Suppose that the preferences for the actions are as listed in Table 3.5. For instance, the information and decision rules available to Al indicate that action 1 is best, followed in turn by actions 0 and – 1.

Table 3.5 Preferences for actions 1, 0 and –1 for three agents.

	Agent		
Rank	Al	Bo	Cy
1	1	0	–1
2	0	–1	1
3	–1	1	0

Let the symbol "$\succ$" indicate preference between two actions. Then we may write $1 \succ 0$ to stipulate that action 1 is preferred to action 0. Using this notation, we can express the entire preference relation for Al as $1 \succ 0 \succ -1$.

If the three decision makers have equal authority, a simple *voting* procedure seems like a reasonable strategy. For the given set of preferences, however, each candidate action would receive exactly one vote. In particular, Al, Bo and Cy would respectively vote for their first choices of 1, 0, and – 1.

What about a process of *veto* to eliminate some or most of the candidate actions? This procedure runs into trouble as well. The coalition {Al, Bo} would vote against option –1, since that option is neither player's first choice. In a similar way, the group {Bo, Cy} would veto action 1, while {Al, Cy} would block the 0 option. In this way, we see that all three candidate actions are blocked.

The vetoing procedure takes account of each agent's first choice, but ignores his relative ranking of the remaining two choices. Can this *preference* information be used in full to generate a reasonable outcome? The answer, again, is negative.

The preference relations for Al, Bo, and Cy are, respectively:

$$1 \succ 0 \succ -1$$
$$0 \succ -1 \succ 1$$
$$-1 \succ 1 \succ 0$$

Is there a preference relation for the group as a whole? By combining the three rankings above into a single expression, we obtain a relation such as

$$1 \succ 0 > -1 \succ 1 \succ 0$$

This would imply, for instance, that $1 \succ 1$. This represents a contradiction, since something cannot be preferred to itself.

In summary, there is no reasonable way to reach a collective decision consistent with the preferences of each individual. A society of autonomous agents is subject to paralysis in this sense. The result is known as Arrow's Impossibility Theorem (Arrow, 1951; Feldman, 1980; Sen, 1986; Sen and Pattanaik, 1969).

Concurrent design

The issue of multiagent decision making arises not only in engineered systems, but in human-level organizations as well. This is illustrated in the need to coordinate collective knowledge among human experts in the concurrent design of a product.

Let $N = \{1, \cdots, n\}$ be the set of agents involved. For instance, 1 may represent the design engineer, 2 the production engineer, and 3 the marketing manager. Each agent will have his own criteria for success, and a corresponding decision function.

Suppose the relevant characteristics of the product are as follows:

P = Performance

M = Manufacturability

K = Unit Cost

The design engineer may have the preference $P \succ M \succ K$; the manufacturing engineer might rank the features as $M \succ P \succ K$; and the marketing manager could well prefer $K \succ P \succ M$. As with the previous examples, there is no clear-cut choice for the organization.

One way to address this problem is to develop, in effect, a relative weighting function. This can be achieved by defining a global utility function for the organization. A plausible function is the gross margin $G = G(P,M,K)$ as a function of the performance P, manufacturability M, and unit cost K. The gross margin is the difference between the revenue R and the total production cost C, each of which depends on P, M and K:

$$G(P,M,K) = R(P,M,K) - C(P,M,K)$$

The precise nature of R and C will depend on the particular application. For instance, if R were linear in P, M and K, we would obtain

$$R(P,M,K) = k_0 + k_1 P + k_2 M + k_3 K$$

Once the nature of R is established, the overall decision problem comes down to the choice of values for P^*, M^* and K^* which maximize the gross margin:

$$G(P^*,M^*,K^*) = \max_{P,M,K} G(P,M,K)$$

Cooperative systems

A *cooperative* system is one where individual agents can combine their efforts to attain their goals. Often these interactions can yield synergism. In other words, the payoff to a group can exceed the sum of the payoffs to the agents acting in isolation.

The existence of synergism is usually the motive force behind a system. If each agent could do as well acting alone as in combination, then there would be little need for their coordination.

A cooperative system may or may not be bound by a common goal. A common goal is found in two people working together to build a single house. On the other hand, two other individuals may cooperate to remove heavy boulders from each of their fields, even though neither has any benign interest in the welfare of the other.

A set of agents united by a common goal is called a *team* (Arrow, 1964; Arrow and Radner, 1979; Chang and Ho, 1983; Marschak, 1955; Marschak and Radner, 1972; Radner, 1962). Teams therefore comprise a subset of cooperative systems.

We first review some critical concepts from cooperative game theory (Owen, 1982; Shubik, 1982; Vorob'ev, 1977). This discussion[2] is followed by a number of applications to the modeling of intelligent action.

Characteristic functions

Consider a finite set of agents $\mathbf{N} = \{1,2,\ldots,n\}$. Any subset $\mathbf{K}$ of this set is called a *coalition*. The largest coalition consists of the entire set $\mathbf{N}$, while the smallest coalition consists of the null set $\emptyset$.

For notational simplicity, we will often refer to a singleton set by its element. In particular, the coalition $\{i\}$ consisting solely of agent i, will usually be denoted simply as i.

Each coalition is associated with one or more values of payoff, depending on the behavior of its members. The maximum payoff for each coalition $\mathbf{K}$ is called its *characteristic function* $v(\mathbf{K})$.

As an example, suppose that $\mathbf{N}$ represents a homogeneous set of workers or machines. Each agent can complete the same amount of work in a day and earn a payoff π. If there is no interaction effect, positive or negative, among the workers, then the payoff to a coalition K is proportional to its numerosity. In other words, $v(\mathbf{K}) = |\mathbf{K}|\,\pi$, where $|\mathbf{K}|$ denotes the cardinality of $\mathbf{K}$.

In a different context, consider a system of three robots, $\mathbf{N} = \{1, 2, 3\}$. Their task is to move a heavy load, for which they receive a net payoff of 1. Suppose that any two robots can cooperate to accomplish the job, while a single agent is inadequate to the task. Then the characteristic function takes a binary form. In particular, $v(\emptyset) = v(1) = v(2) = v(3) = 0$; but $v(\{1, 2\}) = v(\{1, 3\}) = v(\{2, 3\}) = v(\{1, 2, 3\}) = 1$.

Superadditivity. Coalitions tend to be synergistic. More specifically, the payoff to a coalition $\mathbf{K}$ should at least equal the sum of the payoffs to each agent acting alone:

$$v(\mathbf{K}) \geq \sum_{i \in \mathbf{K}} v(i)$$

[2] The discussion of cooperative game theory is adapted from Vorob'ev (1977).

If the situation were otherwise, the agents could, in effect, act independently and merely aggregate their resources.

More generally, any set of coalitions must be synergistic. Suppose $\mathbf{K}_1,\ldots, \mathbf{K}_m$ are disjoint coalitions of the population N; that is, $\mathbf{K}_i \cap \mathbf{K}_j = \emptyset$ for $i \neq j$. Then their union is at least as effective as their component groups acting separately:

$$v\left(\bigcup_{j=1}^{m} \mathbf{K}_j\right) \geq \sum_{j=1}^{m} v(\mathbf{K}_j) \qquad \text{(Eq. 1)}$$

We call this property *synergism* or *superadditivity*.

Additivity. An *additive* game is one in which no interaction effects occur among the agents. Hence the payoff to a coalition **K** is the sum of the payoffs to each member:

$$v(\mathbf{K}) = \sum_{i \in \mathbf{K}} v(i)$$

An additive game may also be called an *isolated* or *noncooperative* game.

The characteristic value of the null coalition $\emptyset$ is zero for an additive game. We can verify this by noting that $\mathbf{K} = \mathbf{K} \cup \emptyset$ and writing

$$v(\mathbf{K}) = v(\mathbf{K} \cup \emptyset) = v(\mathbf{K}) + v(\emptyset)$$

from which we obtain $v(\emptyset) = 0$.

Constant Sum Games and Complementarity. Let $\mathbf{K}_1,\ldots, \mathbf{K}_m$ be any set of coalitions from population N. A *constant sum* game is one where the sum of the payoffs is fixed. In other words,

$$\sum_{j=1}^{m} v(\mathbf{K}_j) = c$$

for some constant c.

In particular, suppose **K** is any coalition and $\overline{\mathbf{K}} \equiv \mathbf{N} - \mathbf{K}$ its complement with respect to the population N. Then for a constant sum game:

$$v(\mathbf{K}) + v(\overline{\mathbf{K}}) = v(\mathbf{N})$$

This is known as the principle of *complementarity*.

Imputations

Let π_i denote the payoff attainable by agent *i*, whether acting alone or in concert with others. These payoffs should satisfy several conditions associated with the nature of the game.

One condition requires that π_i be at least as great as the payoff to individual *i* acting alone. This criterion is known as *individual rationality:*

$$\pi_i \geq v(i)$$

where $v(i)$ is the characteristic value for agent *i*. If the agent were offered less than v(i) in a coalition, he would simply leave and act in isolation.

The payoffs π_i must collectively satisfy a global constraint. In particular, the payoffs must match the return to the population as a whole. This condition is called *group rationality:*

$$\sum_{i \in N} \pi_i = v(N)$$

The preceding relation can be justified as follows. Suppose $\Sigma_i \pi_i < v(N)$. Then the individual agents collectively receive less than the entire payoff available to them. A better alternative exists: the Pareto optimal condition in which at least one agent receives the surplus payoff without causing anyone else to receive less. On the other hand, suppose $\Sigma_i \pi_i > v(N)$. Then the agents are proposing to allocate funds which exceed the total amount available to them: an infeasible situation. By this argument, we conclude that the allocations π_i must collectively match the global budget v(N).

Consider a characteristic function v and a vector $\pi = (\pi_1, \ldots, \pi_n)$ in which the payoffs π_i satisfy the conditions of individual and group rationality. Then π is known as an *imputation* under the characteristic function v.

Let v be a characteristic function for a population **N**. The ordered pair $\langle N, v \rangle$ is called a *classical cooperative* game or a game in *characteristic function form*. (Another type of collaborative game allows for active penalties imposed by agents on each other. For instance, one agent may threaten to hurt another–even at some penalty to the former–in order to secure cooperation.)

A necessary and sufficient condition for the vector $\pi = (\pi_1, \ldots, \pi_n)$ to be an *imputation* is a linear relationship between each element of π_i and the characteristic value $v(i)$. We state this result as a theorem.

Theorem. Consider a classical cooperative game $\langle \mathbf{N}, v \rangle$. Let $(a_1, \ldots, a_n)$ be a vector of nonnegative constants which collectively amount to the difference between the global characteristic value $v(\mathbf{N})$ and the sum of individual characteristic values $v(i)$; that is:

$$\sum_{i \in \mathbf{N}} a_i = v(\mathbf{N}) - \sum_{i \in \mathbf{N}} v(i) \qquad \text{(Eq. 2)}$$

where each $a_i \geq 0$. Then a vector $\pi = (\pi_1, \ldots, \pi_n)$ is an imputation for the game $\langle \mathbf{N}, v \rangle$ if and only if (iff)

$$\pi_i = v(i) + a_i \qquad \text{(Eq. 3)}$$

for each $i \in \mathbf{N}$.

Equation 3 states that the imputed benefit π_i to agent i is linear with respect to the characteristic payoff $v(i)$. Moreover, the utility of joining a coalition must be worth at least as much as independent action. The theorem can be validated as follows.

Proof. The demonstration in the reverse direction is simple. Since $\pi_i \geq v(i)$, individual rationality is satisfied. Further, a few algebraic steps will verify that $\Sigma_i \pi_i = v(\mathbf{N})$, thereby confirming group rationality.

Validation in the forward direction is as follows. Let

$$a_i \equiv \pi_i - v(i) \qquad \text{(Eq. 4)}$$

where the symbol $\equiv$ denotes "defined as". From individual rationality, we can deduce that $a_i \geq 0$. By summing Eq. 4 over all agents, we obtain

$$\Sigma_i \, a_i = \Sigma_i \, \pi_i - \Sigma_i \, v(i) = v(\mathbf{N}) - \Sigma_i \, v(i)$$

The second equality follows from group rationality, which requires $\Sigma_i \pi_i = v(\mathbf{N})$. Thus Eq. 2 is verified. ■

Essential and Inessential Games

Superadditivity of a characteristic function v is a reflection of synergism. When an agent joins a coalition, or two coalitions combine to form a larger group, the resulting payoff to each player is at least as large. In the biological realm, no organ in an animal can survive in isolation but the union of organs in a healthy animal can thrive. The same situation holds for the components of an assembly robot or the functional departments in an enterprise.

Superadditivity in its weakest form is simply an additive relation among coalitions. In algebraic terms,

$$v(\mathbf{K} \cup \mathbf{L}) = v(\mathbf{K}) + v(\mathbf{L})$$

for any two disjoint coalitions **K** and **L** in **N**. When the characteristic function is additive, the agents have no incentive nor any disincentive to join together.

The characteristic function is additive whenever the individual characteristic values $v(i)$ collectively match the group value $v(\mathbf{N})$.

> **Theorem.** A characteristic function v is additive if and only if the global value equals the sum of individual values:

$$v(\mathbf{N}) = \sum_{i \in \mathbf{N}} v(i) \qquad \text{(Eq. 5)}$$

Proof. To show necessity, assume that v is additive. Then Eq. 5 follows immediately.

To verify sufficiency, we first recall that any characteristic function v is superadditive:

$$v(\mathbf{K} \cup \mathbf{L}) \geq v(\mathbf{K}) + v(\mathbf{L})$$

for any two disjoint coalitions **K** and **L**. Let **M** be the complement of **K** and **L**; that is, $\mathbf{M} \equiv \mathbf{N} - (\mathbf{K} \cup \mathbf{L})$. Then $\mathbf{N} = \mathbf{K} \cup \mathbf{L} \cup \mathbf{M}$, and we can write

$$v(\mathbf{N}) = v([\mathbf{K} \cup \mathbf{L}] \cup \mathbf{M}) \geq v(\mathbf{K} \cup \mathbf{L}) + v(\mathbf{M}) \geq v(\mathbf{K}) + v(\mathbf{M}) \qquad \text{(Eq. 6)}$$

The inequalities follow from superadditivity. Using Eq. 5, we obtain

$$v(N) = \sum_{i \in \mathbf{N}} v(i) = \sum_{i \in \mathbf{K}} v(i) = \sum_{i \in \mathbf{L}} v(i) = \sum_{i \in \mathbf{M}} v(i)$$

$$\leq v(\mathbf{K}) + v(\mathbf{L}) + v(\mathbf{M}) \qquad \text{(Eq. 7)}$$

The last expression in Eqs. 6 and 7 are identical. We therefore deduce that all other expressions between the inequalities are equal. In particular,

$$v(\mathbf{K} \cup \mathbf{L}) + v(\mathbf{M}) = v(\mathbf{K}) + v(\mathbf{L}) + v(\mathbf{M})$$

from which we conclude that $v(\mathbf{K} \cup \mathbf{L}) = v(\mathbf{K}) + v(\mathbf{L})$. This ends the sufficiency part of the proof. ■

When a cooperative game has an additive characteristic function, it is said to be *inessential*. All other cooperative games are called *essential*. An essential game reflects synergism[3] among participants; for each agent to receive maximal payoff, it is essential for him to cooperate.

The description of payoffs for an inessential game is simple, as there is only one possibility. On the other hand, an essential game has infinitely many. This result is stated as a theorem.

Theorem. An inessential game has precisely one imputation. In particular, the imputation is

$$(v(1),\ldots, v(n))$$

In contrast, an essential game with two or more agents has an infinite number of imputations.

Proof. From Eq. 3, we can write the imputation for agent i as $\pi_i = v(i) + a_i$, where $a_i \geq 0$. By summing over all agents, we obtain

$$\textstyle\sum_i \pi_i = \sum_i v(i) + \sum_i a_i \qquad \text{(Eq. 8)}$$

From group rationality, $\Sigma_i \pi_i = v(N)$. Since the game is inessential, the characteristic function is additive and we can write $\Sigma_i v(i) = v(N)$. Hence Eq. 8 yields $\Sigma_i a_i = 0$. But the a_i are nonnegative. The last two statements imply that $a_i = 0$. Consequently, the imputation vector becomes $(\pi_1, \ldots, \pi_n) = (v(1),\ldots,v(n))$.

In contrast, an essential game exhibits synergism. From Eq. 8, we can write

$$\textstyle\sum_i a_i = \sum_i \pi_i - \sum_i v(i) = v(N) - \sum_i V(i) > 0$$

The second equality follows from group rationality, and the inequality from synergism. The positive amount $\Sigma_i a_i > 0$ can be partitioned among the a_i parameters in the imputations $\pi_i = v(i) + a_i$ in an infinite number of ways. ■

Strategic Equivalence

Cooperative games can be classified according to their properties into different categories. One such classification takes the form of linear relationships among characteristic functions. A linear relation between two characteristic functions implies a similarity in the consequences for their underlying games.

---

[3] Synergism is defined by non-additivity in the characteristic value. Consequently, it may lead to inconvenience in practical terms, due to the computational burden.

Consider a cooperative game involving a set **N** of agents. Let v and v' be two characteristic functions. The game $\Gamma = \langle \mathbf{N}, v\rangle$ is said to be *strategically equivalent* to the game $\Gamma' = \langle \mathbf{N}, v'\rangle$ if the following relationship holds:

$$v'(\mathbf{K}) = kv(\mathbf{K}) + \sum_{i \in \mathbf{K}} c_i \qquad \text{(Eq. 9)}$$

where k is a positive number, each c_i is any real number, and **K** is a subset of **N**.

The strategic equivalence of cooperative games is defined in terms of their characteristic functions. For this reason, we also speak of the strategic equivalence of the characteristic functions rather than only of their underlying games.

The constant term in Eq. 9 consists of the sum of values c_i associated with each agent i. The magnitude of c_i represents the fixed payoff to agent i for playing the game. A positive value for c_i indicates a profit; and a negative value, a loss. Consequently the sum given by $\Sigma_i c_i$ for $i \in \mathbf{K}$ represents the collective payment to the agents in the coalition **K** simply for participating in the game.

On the other hand, the scaling factor k reflects the unit of currency; or more precisely, the exchange rate between the currencies used for characteristic functions v and v'. The value of k may take the form of so many dollars per franc; or x watt-hours per ruble; or even 5 dollars per 2 dollars, reflecting the scale of compensation.

In summary, $\Sigma_i c_i$ for $i \in \mathbf{K}$ is a constant of displacement while k is a constant of scale. The strategic implications for two games should be unaffected by their cost of participation and unit of reward. This is the motivation behind the term "strategic equivalence."

We will use the symbol $\approx$ to denote strategic equivalence. When two games $\Gamma = \langle \mathbf{N}, v\rangle$ and $\Gamma' = \langle \mathbf{N}, v'\rangle$ are strategically equivalent, we will write $\Gamma \approx \Gamma'$ or simply $v \approx v'$.

Properties of strategic equivalence. Strategic equivalence is a relation which satisfies the properties of reflexivity, symmetry, and transitivity.

- *Reflexivity.* A characteristic function is strategically equivalent to itself:

  $$v \approx v$$

  This can be shown by setting $k = 1$ and each $c_i = 0$ for each $i \in N$ in Eq. 9.

- *Symmetry.* If a characteristic function v is strategically equivalent to another characteristic function v', then the converse is also true:

  $$(v \approx v') \rightarrow (v' \approx v)$$

To verify this property, we first assume that Eq. 9 holds. Then we can write

$$v(\mathbf{K}) = \frac{1}{k} v'(\mathbf{K}) + \sum_{i \in \mathbf{K}} \left(\frac{-c_i}{k} \right)$$

We define new constants $k' \equiv 1/k$ and $c_i' \equiv -c_i/k$ to obtain

$$v(\mathbf{K}) = k'v'(\mathbf{K}) + \sum_{i \in \mathbf{K}} c_i'$$

Since k is positive, so is k'. Hence v' is strategically equivalent to v.

- *Transitivity.* Suppose v is strategically equivalent to v', and v' to v''; then v is strategically equivalent to v'':

$$(v \approx v') \,\&\, (v' \approx v'') \rightarrow (v \approx v'')$$

Based on the assumptions, the following conditions hold:

$$v'(\mathbf{K}) = kv(\mathbf{K}) + \sum_{i \in \mathbf{K}} c_i$$

$$v''(\mathbf{K}) = k'v'(\mathbf{K}) + \sum_{i \in \mathbf{K}} c_i'$$

where both k and k' are positive. By substituting the first equation into the second, we can write

$$v''(\mathbf{K}) = \left[k'kv(\mathbf{K}) + \sum_{i \in \mathbf{K}} kc_i \right] + \sum_{i \in \mathbf{K}} c_i'$$

We now define $k'' \equiv k'k$ and $c_i'' \equiv kc_i + c_i'$ and obtain

$$v''(\mathbf{K}) = k''v(\mathbf{K}) + \sum_{i \in \mathbf{K}} c_i''$$

Since both k and k' are positive, so is k''. Thus v is strategically equivalent to v''.

We have just seen that strategic equivalence exhibits the properties of reflexivity, symmetry, and transitivity. These properties imply a partition of the set of all characteristic functions. In other words, characteristic functions can be subdivided into distinct classes.

Strategic equivalence of imputations. Strategic equivalence between games generates a strategic equivalence between their imputations. In particular, let v and v′ be strategically equivalent:

$$v'(\mathbf{K}) = kv(\mathbf{K}) + \sum_{i \in \mathbf{K}} c_i$$

Further, suppose that $\pi = (\pi_1,\ldots,\pi_n)$ is an imputation for v. Then we can easily construct an imputation for v' in the following way.

We define the payoff π_i' for agent i under the characteristic function v' by a linear transformation on π_i. In particular, let $\pi_i' \equiv k\pi_i + c_i$. Due to individual rationality for v, we know that $\pi_i \geq v(i)$. Hence

$$\pi_i' \equiv k\pi + c_i \geq kv(i) + c_i = v'(i)$$

The last equality follows from setting $\mathbf{K} = \{i\}$ in Eq. 9. Since $\pi_i' \geq v'(i)$ for each i, individual rationality is satisfied for the vector $\pi' \equiv (\pi_1',\ldots,\pi_n')$.

By summing over all agents, we can write

$$\sum_{i \in \mathbf{N}} \pi'_i = \sum_{i \in \mathbf{N}} (k\pi_i + c_i) = k \sum_{i \in \mathbf{N}} \pi_i + \sum_{i \in \mathbf{N}} c_i$$

$$= kv(\mathbf{N}) + \sum_{i \in \mathbf{N}} c_i = v'(\mathbf{N})$$

The third equality follows from group rationality for π, while the last equation relies on the strategic equivalence of v and v'. Since $\Sigma_{i \in \mathbf{N}}\ \pi_i' = v'(\mathbf{N})$, group rationality is satisfied for v′. Hence the vector π' is an imputation for the game $\langle \mathbf{N}, v' \rangle$.

Zero Game

In this section, we define a game with the simplest possible characteristic function. We will see that every inessential game is reducible to this canonical form in a strategic sense.

A cooperative game $\langle \mathbf{N}, v \rangle$ is called a *zero game* if the characteristic function v is zero for every coalition. A zero game is bland in the sense that no agent receives either a reward or penalty for participation.

An interesting result is the strategic similarity of the zero game to all inessential games.

Theorem. Every inessential game is equivalent to a zero game.

Proof. We recall that an inessential game $\langle \mathbf{N},v\rangle$ has an additive characteristic function:

$$\sum_{i\in\mathbf{K}} v(i) = v(\mathbf{K})$$

By choosing $k \equiv 1$ and $c_i \equiv -v(i)$ in Eq. 9, we obtain

$$v'(\mathbf{K}) = v(\mathbf{K}) - \sum_{i\in\mathbf{K}} v(i)$$

for any coalition **K**. The last two relations imply that $v'(\mathbf{K}) = 0$. Hence the game $\langle \mathbf{N},v\rangle$ is strategically equivalent to the zero game. ■

In consequence, the set of all inessential games $\langle \mathbf{N},v\rangle$ comprises a single class of strategic equivalence–namely, that of a zero game.

Binary Game

Almost as simple as a zero game is one whose characteristic function takes a binary form. We will see that each essential cooperative game corresponds to precisely one binary game.

A cooperative game $\langle \mathbf{N},v\rangle$ has a *binary form* if its characteristic function is unity for the entire group, and zero for individuals:

$$v(i) = 0 \quad : \quad i \in \mathbf{N}$$
$$v(N) = 1$$

Essential games have their counterparts in binary form.

Theorem (Binary Reduction). Each essential cooperative game is strategically equivalent to precisely one game in binary form.

Proof. Let $\langle \mathbf{N},v\rangle$ be an essential cooperative game. We define the characteristic function v' by the following relations:

$$v'(i) = kv(i) + c_i = 0 \quad : i = 1,\cdots, n \qquad \text{(Eq. 10)}$$

$$v'(\mathbf{N}) = kv(\mathbf{N}) + \sum_{i\in\mathbf{N}} c_i = 1 \qquad \text{(Eq. 11)}$$

These statements define $n + 1$ equations for the $n + 1$ unknown quantities $\{c_1, \ldots, c_n\}$ and k.

Summing Eq. 10 over all n agents yields

$$k \sum_{i \in \mathbf{N}} v(i) + \sum_{i \in \mathbf{N}} c_i = 0$$

Inserting the result into Eq. 11 yields

$$k \left[v(\mathbf{N}) - \sum_{i \in \mathbf{N}} V(i) \right] = 1$$

which can be written as

$$k = \frac{1}{v(\mathbf{N}) - \sum_{i \in \mathbf{N}} v(i)}$$

The denominator must be positive, since the characteristic function v defines an essential game. Thus k is also positive. Returning to Eq. 10, we have

$$_i c = -kv(i) = \frac{-v(i)}{v(\mathbf{N}) - \sum_{i \in \mathbf{N}} v(i)}$$

for each i.

These expressions for the constants k and c_i uniquely specify the values of $v'(i)$ from $v(i)$ in Eq. 10. Since $v'(i) = 0$ and $v'(\mathbf{N}) = 1$, the characteristic function v' defines a game in binary form. Hence the essential game $\langle \mathbf{N}, v \rangle$ corresponds to a unique binary game.

In addition to binary forms, it is possible to consider games in "*a:b*" reduced form. Here a and b are any two constants for which $na \neq b$. Games in this form are defined by the relations

$$v'(i) = a \quad : \quad i = 1, \ldots, n$$
$$v'(\mathbf{N}) = b$$

It is straightforward to show that any inessential game can be reduced to a game in *a:b* form. The reasoning process is similar to that for reduction to binary form.

In particular, payoffs of the form " −1 : 0" often arise in games which model competitive behavior. As we have already discussed, however, all these forms correspond strategically to each other.

Geometric Interpretation

Let $\pi = (\pi_1, \ldots, \pi_n)$ be any vector whose elements define a convex combination of weights. In other words,

$$\pi_i \geq 0 \quad : i = 1, \ldots, n$$

$$\sum_{i=1}^{n} \pi_i = 1$$

These values would satisfy the conditions of individual rationality:

$$\pi_i \geq v(i) = 0$$

and group rationality:

$$\sum_{i=1}^{n} \pi_i = \sum_{i \in N} v(i) = 1$$

for a game in binary form. Consequently, such a vector π can represent an imputation for a game in binary form. Since the elements $\pi_1, \ldots, \pi_n$ represent a convex combination, each π_i can be regarded as the payoff to agent i relative to the others.

In geometric terms, the unit vectors in n dimensions collectively define a hyperplane of dimension $n - 1$ through the equation

$$\pi_1 + \cdots + \pi_n = 1$$

The convex hull of these unit vectors, known as the fundamental simplex, defines the space of possible payoffs.

The case for 3 dimensions is portrayed in Figure 3.2. The fundamental simplex is the triangle defined by the convex combination of the unit vectors (1,0,0), (0,1,0) and (0,0,1). The imputation $\pi = (\frac{1}{3}, \frac{1}{3}, \frac{1}{3})$, lying at the center of the triangle, represents an equality of payoffs to all three agents. In contrast, the apex of the triangle represented by $\pi = (0,0,1)$ corresponds to the situation in which agent 3 receives the entire payoff.

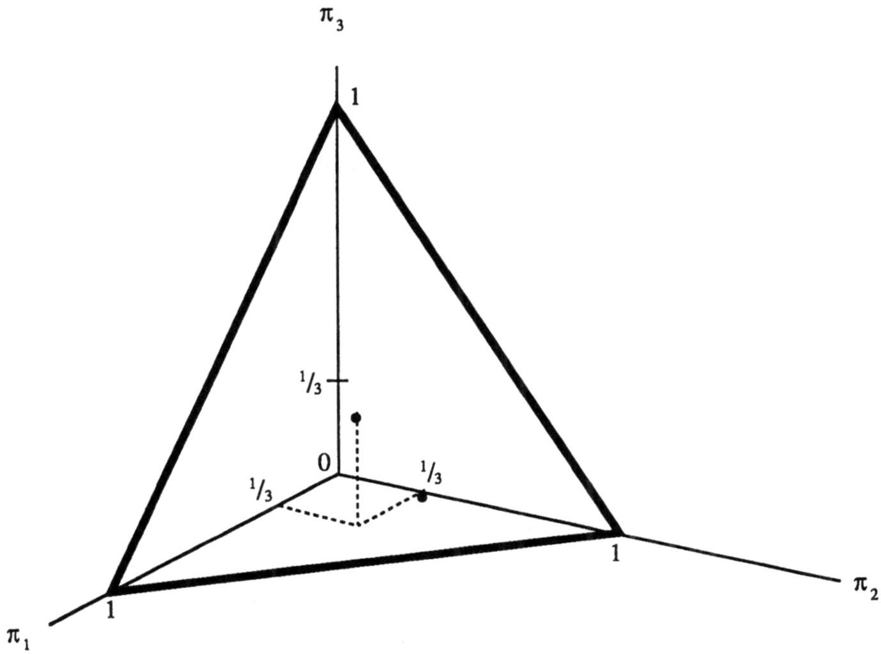

Figure 3.2 Fundamental simplex in 3 dimensions is a triangle. The imputation (1/3, 1/3, 1/3) lies at the center of the triangle.

Games with Few Players

Constant-sum games with only two or three players are strategically equivalent to simpler games. For instance, all constant sum games between two agents are inessential.

Theorem. Any cooperative constant sum game involving two agents is inessential.

Proof. Suppose that an essential cooperative game $\langle \mathbf{N}, v \rangle$ exists. By the Binary Reduction Theorem, this game is strategically equivalent to a binary game with a characteristic function v. The binary game is defined by the relations

$$v(1) = 0$$
$$v(2) = 0$$
$$v(\{1,2\}) = 1$$

On the other hand, the definition of a constant-sum game requires that all partitions of the population N yield the same characteristic value. In particular, we require the property of complementarity: if $\mathbf{K}$ is any subset of $\mathbf{N}$, then

$$v(\mathbf{N} - \mathbf{K}) = v(\mathbf{N}) - v(\mathbf{K})$$

In our case, letting $\mathbf{K} = \{1\}$ yields

$$v(2) = v(\{1,2\}) - v(1)$$

By replacing these three terms with their values from the preceding paragraph, we obtain 0 = 1. This contradiction implies that a constant-sum game cannot be essential.■

We now turn to cooperative games involving three agents. We have seen in an earlier section that all inessential games comprise a single class of strategic equivalence–namely, that of the zero game.

How about essential, 3-agent games? Let us consider these games in binary form. This reduced form has the following characteristic values:

$$v(1) = v(2) = v(3) = 0$$
$$v(\{1,2,3\}) = 1$$

The property of complementarity implies

$$v(\{1, 2\}) = v(\{1, 2, 3\}) - v(3) = 1 - 0 = 1$$

In a similar way, we deduce that $v(\{1, 3\})$ and $v(\{2, 3\})$ are both equal to 1. These values uniquely specify the characteristic function for 3-agent, essential games.

To summarize, cooperative games involving 3 players fall into two classes of strategic equivalence. One category involves inessential games, which belong to the class of the zero game. The second category consists of essential games whose common reduced form is

$$v(\mathbf{K}) = \begin{cases} 0 : |\mathbf{K}| \leq 1 \\ 1 : |\mathbf{K}| \geq 2 \end{cases}$$

for any $\mathbf{K} \subset \mathbf{N}$.

Games involving four agents exhibit a richer behavior space, as would be expected. The simplest category involving four players is again the class of inessential games, which matches the class of the zero game.

We now turn to the binary forms for essential games. Let $\langle \mathbf{N}, v' \rangle$ be an essential 4-agent game with the binary form $\langle \mathbf{N}, v \rangle$. Then we know that

$$v(1) = v(2) = v(3) = v(4) = 0$$
$$v(\{1,2,3,4\}) = 1$$

The property of complementarity implies that

$$v(\{1,2,3\}) = v(\{1,2,3,4\}) - v(\{4\}) = 1 - 0 = 1$$

By similar reasoning, we deduce that

$$v(\{1,2,4\}) = v(\{1,3,4\}) = v(\{2,3,4\}) = 1$$

What remains for us is to describe the characteristic function for all coalitions containing two agents. We can identify six such coalitions. Due to the complementarity property, however, these coalitions are constrained by three relations:

$$v(\{1, 2\}) = v(\{1, 2, 3, 4\}) - v(\{3, 4\}) = 1 - v(\{3, 4\})$$
$$v(\{1, 3\}) = 1 - v(\{2, 4\})$$
$$v(\{1, 4\}) = 1 - v(\{2, 3\})$$

Six variables in three equations imply three degrees of freedom.

Each of these variables, however, is also subject to lower and upper bounds. A coalition of two agents must be worth at least as much as each acting in isolation; since the value of the latter is 0, this is a lower bound on the worth of the coalition. In the opposite direction, a coalition of two members cannot be worth more than an alliance of three agents; since the value of the latter is 1, this represents an upper limit on each 2-agent coalition.

Since there are three degrees of freedom, let us define three new variables:

$$x_1 \equiv v(\{1,2\})$$
$$x_2 \equiv v(\{1,3\})$$
$$x_3 \equiv v(\{1,4\})$$

The remaining 2-person coalitions can be defined in terms of these variables:

$$v(\{2, 3\}) = 1 - x_3$$
$$v(\{2, 4\}) = 1 - x_2$$
$$v(\{3, 4\}) = 1 - x_1$$

The geometric interpretation of (x_1, x_2, x_3) is the unit cube in 3-dimensional space, as shown in Figure 3.3. Each point in this unit cube represents a class of strategic equivalence. For instance, the point $P = (1, 1, 0.5)$ represents the characteristic function for the first three couples:

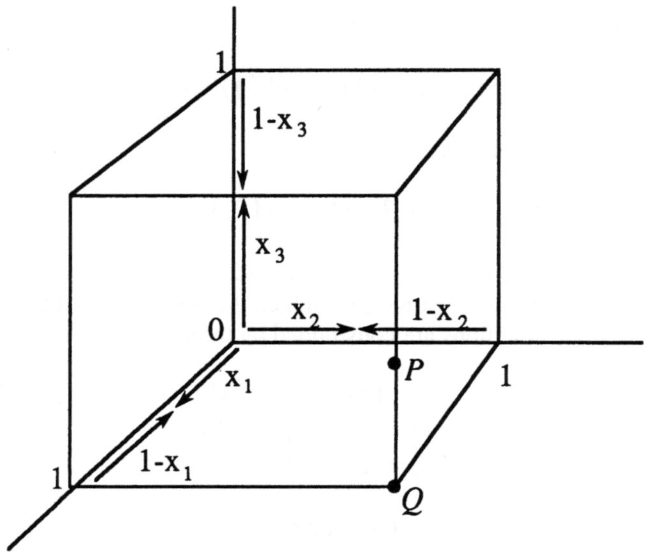

Figure 3.3 Degrees of freedom for essential, 3-agent games. For instance, the point Q=(1,1,0) represents the strategic class whose binary form is x_1=v({1,2})=1, x_2=v({1,3})=1, and x_3=v({1,4})=0. The other 2-member coalitions are associated with the complementary values.

$$v(\{1, 2\}) \equiv x_1 = 1$$
$$v(\{1, 3\}) \equiv x_2 = 1$$
$$v(\{1, 4\}) \equiv x_3 = 0.5$$

The remaining 2-agent coalitions are defined by their respective complements:

$$v(\{2, 3\}) \equiv 1 - x_3 = 0$$
$$v(\{2, 4\}) \equiv 1 - x_2 = 0$$
$$v(\{3, 4\}) \equiv 1 - x_1 = 0.5$$

Since each point (x_1, x_2, x_3) in the unit cube defines a category of strategic equivalence, the number of strategic classes for essential, 4-agent, constant-sum games is infinite. This is a remarkable jump in complexity from the case of two- or three-agent games.

The degrees of freedom for cooperative, constant-sum games expands combinatorially with the number of agents. A game with 5 agents has 10 degrees of freedom for the number of strategic classes. More generally, a game with n agents results in $2^{n-1} - n - 1$ degrees of freedom (Vorob'ev, 1977, p. 131).

Nonconstant-Sum Games. Let us consider nonconstant-sum games. The case of two agents is simple. The binary form is defined by the following characteristic function:

$$v(1) = v(2) = 0$$
$$v(\{1, 2\}) = 1$$

These values uniquely characterize every such game. In other words, there is only a simple strategic class of essential, 2-agent cooperative games.

For 3-agent games, the binary reduction exhibits the following boundary points:

$$v(1) = v(2) = v(3) = 0$$
$$v(\{1, 2, 3\}) = 1$$

A coalition containing two members can take any value between the lower limit of 0 and the upper limit of 1, including the boundary points.

Since the game is nonconstant-sum, there are no other constraints. Hence we may define three new variables:

$$x_1 \equiv w(\{1, 2\})$$
$$x_2 \equiv w(\{1, 3\})$$
$$x_3 \equiv w(\{2, 3\})$$

These variables imply three degrees of freedom.

Since each x_i lies in the interval [0, 1], the vector (x_1, x_2, x_3) defines a unit cube in 3-dimensional space. Consequently, the variety of strategic classes for 3-agent, nonconstant-sum games is identical to that of 4-agent, constant-sum games.

This result is intuitively appealing. The imposition of a constant-sum constraint on a game of n agents reduces the dimensionality of its characteristic function by 1. Hence the degrees of freedom for a constant-sum game involving n agents matches the dimensionality of nonconstant-sum games for $n - 1$ agents.

Dominance of Imputations

An imputation refers to the allocation of the overall payoff to a group among its members. So far we have examined some necessary properties of imputations, but have said nothing about preference orderings among them. Given two distinct imputations, the one which offers a higher payoff to a particular agent will look attractive to the beneficiary, and less so to another agent who would gain more from the other imputation.

We first define some terms relating to the relative attractiveness of imputations. This is followed by the relationship among imputations for games within a particular class of strategic equivalence.

Consider the cooperative game $\Gamma = \langle \mathbf{N}, v \rangle$. Let $\mathbf{K} \subset \mathbf{N}$ be a coalition in the game, while $\pi = (\pi_1, \ldots, \pi_n)$ and $\rho = (\rho_1, \ldots, \rho_n)$ are two imputations.

We say that imputation π *dominates* imputation ρ *with respect to coalition* **K** if the following conditions hold:

- *Feasibility.* The collective payoffs do not exceed the return to the coalition:

$$\sum_{i \in \mathbf{K}} \pi_i \leq v(\mathbf{K})$$

- *Preference.* Each payoff in π exceeds that in ρ:

$$\pi_i > \rho_i \text{ for all } i \in \mathbf{K} \qquad \text{(Eq. 12)}$$

The symbol "$\succ_{\mathbf{K}}$" will be used to denote dominance with respect to coalition **K**. In particular, $\pi \succ_{\mathbf{K}} \rho$ indicates that the imputation π dominates ρ for coalition **K**.

The first condition in the preceding definition requires that the proposed allocation should not exceed the income available to the group as a whole. The second condition requires unanimous preference: each agent in a coalition must benefit in order for one imputation to dominate another. This is a strong requirement for the attractiveness of an imputation.

The previous definition qualified dominance in terms of a particular coalition **K**. From a larger perspective, we say that an imputation π *dominates* another imputation ρ if some such coalition can be found. In other words, imputation π dominates ρ if there exists a coalition **K** for which $\pi \succ_{\mathbf{K}} \rho$. In that case, we can also write $\pi \succ \rho$.

What are some properties of the dominance relation? Due to the strict inequality in Eq. 12, dominance does not have the property of reflexivity.

In general, symmetry is not an inherent property of the dominance relation. To show that symmetry may or may not hold, consider a particular coalition **K** and its complement given by $\overline{\mathbf{K}} \equiv \mathbf{N} - \mathbf{K}$. Let π dominate ρ through coalition **K**. Since $\pi_i > \rho_i$ for each i in **K**, we have

$$\sum_{i \in \mathbf{K}} \pi_i > \sum_{i \in \mathbf{K}} \rho_i$$

From the criterion of group rationality, the elements of an imputation must sum to unity:

$$1 = \sum_{i\in \mathbf{K}} \pi_i + \sum_{i\in \overline{\mathbf{K}}} \pi_i$$

combining the previous two relations yields

$$\sum_{i\in \overline{\mathbf{K}}} \pi_i = 1 - \sum_{i\in \mathbf{K}} \pi_i < 1 - \sum_{i\in \mathbf{K}} \rho_i = \sum_{i\in \overline{\mathbf{K}}} \rho_i$$

The first term represents the total payoff available to the "countercoalition" $\overline{\mathbf{K}}$ through the imputation π. Similarly, the last term specifies the collective amount available to $\overline{\mathbf{K}}$ under imputation ρ. The difference given by

$$\Delta \equiv \sum_{i\in \overline{\mathbf{K}}} \rho_i - \sum_{i\in \overline{\mathbf{K}}} \pi_i > 0$$

is the bonus attributable to imputation ρ over π for the countercoalition. Suppose that each $\rho_i > \pi_i$ for $\overline{\mathbf{K}}$; then ρ dominates π for this alliance, and we can write $\rho \succ \pi$. We have $\Sigma_i \rho_i > \Sigma_i \pi_i$ for $i \in \overline{\mathbf{K}}$; however, there might be some value ρ_i which is less than π_i. Then ρ cannot dominate π.

No general argument can be offered for other potential coalitions and their consequences for the dominance relation. Hence symmetry may hold for a particular pair of imputations or not, depending on the circumstances.

In general, transitivity is also not a requisite property of the dominance relation. For a specific example, consider a 4-agent game with the imputations $\pi = \sigma = (0.4, 0.4, 0.1, 0.1)$ and $\rho = (0.3, 0.3, 0.2, 0.2)$. Given the relations

$$\pi_1 = \pi_2 = 0.4 > 0.3 = \rho_1 = \rho_2$$

imputation π dominates ρ for the coalition {1,2}. Hence $\pi \succ \rho$. Further, a comparison of ρ and σ yields

$$\rho_3 = \rho_4 = 0.2 > 0.1 = \sigma_3 = \sigma_4$$

So ρ dominates σ for the coalition {3,4}, allowing us to write $\rho \succ \sigma$. In short, π dominates ρ, which in turn domintes σ. Since $\pi = \sigma$, however, π cannot dominate σ for any coalition. Consequently transitivity does not hold in this case.

Suppose we introduce the imputation $\tau = (0.2, 0.2, 0.3, 0.3)$. For the coalition {1, 2}, π dominates ρ which in turn dominates τ; in addition, π dominates τ. Hence transitivity holds in this case.

In general, then, transitivity is not an inherent property of the dominance relation. Rather, it depends on the particular application at hand.

The recent discussion has focused on general properties – or lack of such – for the dominance relation. Now we turn to some specific coalitions which are incapable of yielding a dominance relation. These coalitions take the form of singletons or the entire population.

> **Theorem.** Neither a singleton nor the entire population can give rise to a dominance relation.

Proof. We first consider a singleton $\{i\}$ for the game $\langle N,v\rangle$. Let π and ρ be two imputations. Suppose π were to dominate ρ with respect to the coalition $\{i\}$. Then the preference condition stipulates that $\pi_i > \rho_i$.

Now the feasibility condition for domination requires $\pi_i \leq v(i)$. Combining these two relations yields

$$\rho_i < \pi_i \leq V(i)$$

A consequence is $\rho_i < v(i)$, which would contradict the requirement of individual rationality for imputation ρ. Hence no singleton can generate a dominance relation.

We next consider the entire population N. Suppose π dominates ρ for N. This implies $\pi_i > \rho_i$ for every i. By summing over all agents, we obtain

$$\sum_{i\in N} \pi_i > \sum_{i\in N} \rho_i = v(N)$$

The equality above relies on the criterion of group rationality for imputation ρ. However, the relation $\Sigma_{i\in N}\ \pi_i > v(N)$ violates group rationality for π. Consequently, the population in its entirety cannot support a dominance relation. ■

Strategic invariance of dominance. The relation of dominance is preserved among games within a particular strategic class.

> **Theorem** (Strategic Invariance). Let $\langle N,v\rangle$ and $\langle N,v'\rangle$ be two games belonging to the same class of strategic equivalence. Suppose π and ρ are imputations pertaining respectively to v and v', while π' and ρ' are the corresponding imputations under the strategic equivalence. Then dominance with respect to a coalition $K\subset N$ is invariant under the transformation; that is,

$$\pi \succ_K \rho \rightarrow \pi' \succ_K \rho'$$

Proof. We assume that v and v' are strategically equivalent. Then there is some constant $k > 0$ for which

$$v'(\mathbf{K}) = kv(\mathbf{K}) + \sum_{i \in \mathbf{K}} c_i$$

where **K** is any coalition in **N**. The dominance condition $\pi \succ_{\mathbf{K}} \rho$ can be specified explicitly in terms of feasibility and preference relations:

$$\sum_{i \in \mathbf{K}} \pi_i \leq v(\mathbf{K}) \qquad \text{(Eq. 13)}$$

$$\pi_i > \rho_i \quad : i \in \mathbf{N} \qquad \text{(Eq. 14)}$$

The imputation π_i' corresponding to π_i under strategic equivalence is given by $\pi_i' = k\pi_i + c_i$. By summing over the coalition, we obtain

$$\sum_{i \in \mathbf{K}} \pi_i' = \sum_{i \in \mathbf{K}} [k\pi_i + c_i] = k \sum_{i \in \mathbf{K}} \pi_i + \sum_{i \in \mathbf{K}} c_i$$

$$\leq kv(\mathbf{K}) + \sum_{i \in \mathbf{K}} c_i \equiv v'(\mathbf{K})$$

The inequality depends on Eq. 13. A consequence of Eq. 14 is

$$\pi_i' \equiv k\pi_i + c_i > k\rho_i + c_i \equiv \rho_i'$$

The last couple of results verify the feasibility of π' as well as its preference over ρ' for the coalition. In other words, $\pi' \succ_{\mathbf{K}} \rho'$. ■

This theorem implies that the relationships concerning the dominance relation carry across all games within a class of strategic equivalence. As a result, an argument made for one game in a strategic class applies to all its siblings.

More specifically, the dominance relation can be explored for all inessential games using the zero game as a canonical form. In addition, dominance for essential games can be examined by focusing on games in the binary form as exemplars.

Dominance for Strategic Classes. We have seen that inessential games correspond to the zero game, which has precisely one imputation. Consequently, the dominance relation cannot arise for such games.

We now turn to essential, constant-sum games involving 3 agents. From the preceding theorem, it is sufficient to consider such games in their binary form.

Suppose that π and ρ are two imputations. From an earlier discussion, we know that dominance cannot arise from any singleton $\{i\}$ nor the entire population $\{1, 2, 3\}$. Consequently we need to consider only the dual coalitions $\{1,2\}$, $\{1,3\}$ and $\{2,3\}$.

We have seen that any imputation can be regarded as a point in the fundamental simplex defined by unit vectors of the same dimension. For three dimensions, the fundamental simplex takes the form of a triangle, as shown in Figure 3.2. An imputation $\pi = (\pi_1,\pi_2,\pi_3)$ is illustrated geometrically in Figure 3.4. Each element π_i of the imputation π represents the distance of the point from the side opposite the ith vertex.

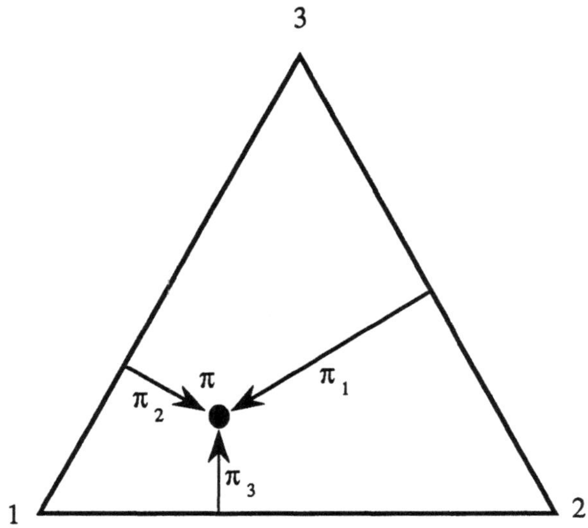

Figure 3.4 An imputation as a point on a triangle. The imputation $\pi=(\pi_1,\pi_2,\pi_3)$ consists of elements π_1 which define the distance of the point from the side opposite vertex i. For instance, π_1 is the distnace from the side 2-3.

Suppose that the coalition $\{1,2\}$ prefers π to ρ. The conditions of domination require feasibility and preference:

$$\begin{aligned} \pi_1 + \pi_2 &\le v(\{1,2\}) \\ \pi_1 &> \rho_1 \\ \pi_2 &> \rho_2 \end{aligned} \qquad \text{(Eq. 15)}$$

From the complementarity property of constant-sum games,

$$v(\{1,2\}) = v(\{1,2,3\}) - v(3) = 1 - 0 = 1$$

On the other hand, $\pi = (\pi_1, \pi_2, \pi_3)$ represents a convex combination of weights; hence we can write

$$\pi_1 + \pi_2 = 1 - \pi_3 \leq 1$$

These results imply that the first relation in Eq. 15 is satisfied.

The conditions $\pi_i > \rho_i$ in Eq. 15 require that the point π lie closer to vertex i than the point ρ. In particular, $\pi_1 > \rho_1$ stipulates the relative proximity of π to vertex 1. This means that ρ must lie to the northeast of the line segment AB, which is parallel to the segment 2-3 in Figure 3.5. In a similar way, the condition $\pi_2 > \rho_2$ requires that ρ lie to the northwest of line segment CD, which is parallel to side 1-3 of the triangle. The two conditions collectively imply that ρ lies above π, within the parallelogram $B\pi D3$. (Conversely, we may state that π must lie below ρ, within the triangle $E\rho F$.) The shaded area in the figure represents the feasible regions for ρ under the domination $\pi \succ_{\{1,2\}} \rho$ with respect to coalition {1,2}. The boundary of the parallelogram is half open: the top portion given by the arc $B3D$ lies in the feasible region. On the other hand, the bottom portion $B\pi D$ is excluded, due to the strict inequality in $\pi_i > \rho_i$ for $i = 1, 2$.

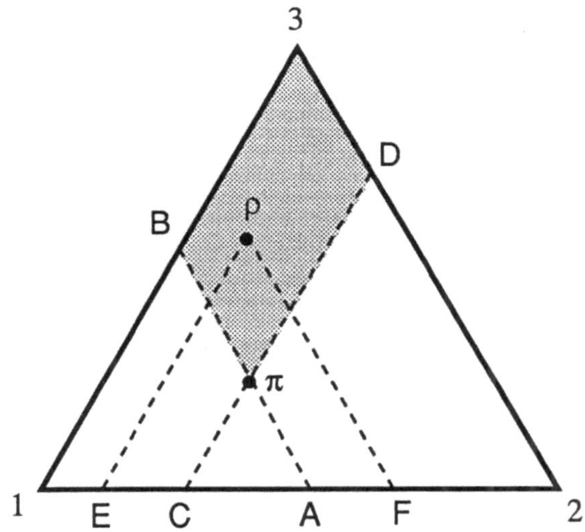

Figure 3.5 The relation $\pi_1 > \rho_1$, requires that ρ lie to the northeast of line segment AB. Similarly, $\pi_2 > \rho_2$ calls for ρ lying to the northwest of segment CD. These two conditions collectively require that ρ lie within the parallelogram $B\pi D3$. The feasible region is half-open: the curve B3D belongs to the set, but the curve $B\pi D$ does not.

The allowable regions due to the coalitions {1,3} and {2,3} can be obtained in similar ways. In particular, the feasible set for {1,3} is the parallelogram touching the vertex 2, while that for {2, 3} includes vertex 1. As a result, the shaded areas in Figure 3.6 depict all the imputations dominated by imputation π for an essential 3-agent, constant-sum game.

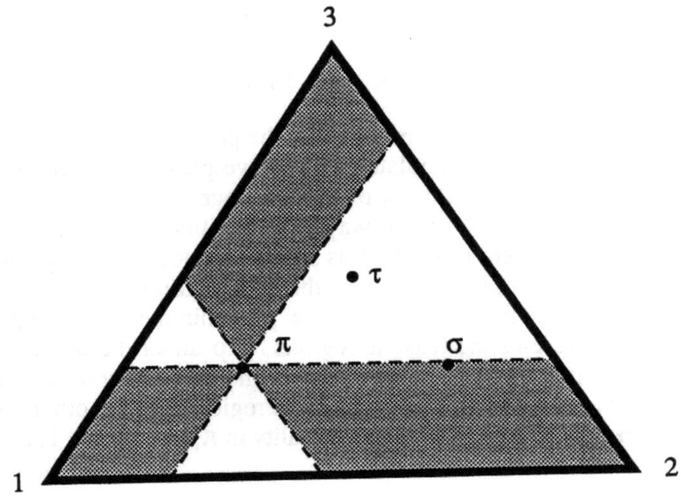

Figure 3.6 Shaded regions depict all the imputations dominated by imputation π. On the other hand, π is dominated by τ. For imputations π and σ, neither dominates the other.

The white regions in Figure 3.6 represent the points which can dominate imputation π. For instance, the point τ dominates π (through the coalition {2, 3}). The only way for each of two points to avoid dominating the other is for them to lie on a line parallel to a side of the triangle. As an illustration, neither of the points π and σ in the figure dominates the other.

We now turn to general 3-agent games; these games might be constant-sum games or not. Without loss of generality, we assume that the game $\langle \{1, 2, 3\}, v \rangle$ is in binary form.

Suppose that imputation π dominates ρ through coalition {1,2}. Since the game is nonconstant-sum, the characteristic function $v(\{1, 2\})$ may take any value in the interval [0, 1]; we denote this value as c_3. We can write

$$\begin{aligned} \pi_1 + + \pi_2 \leq v(\{1, 2\}) \equiv c_3 \\ \pi_1 > \pi_2 \\ \pi_2 > \rho_2 \end{aligned} \qquad \text{(Eq. 16)}$$

The weights π_i sum to unity. Hence the first statement in Eq. 16 can be written as

$$\pi_1 + \pi_2 = 1 - \pi_3 \leq c_3$$

A consequence is

$$\pi_3 \geq 1 - c_3$$

The feasible region is a shaded triangle, as shown in Figure 3.7.

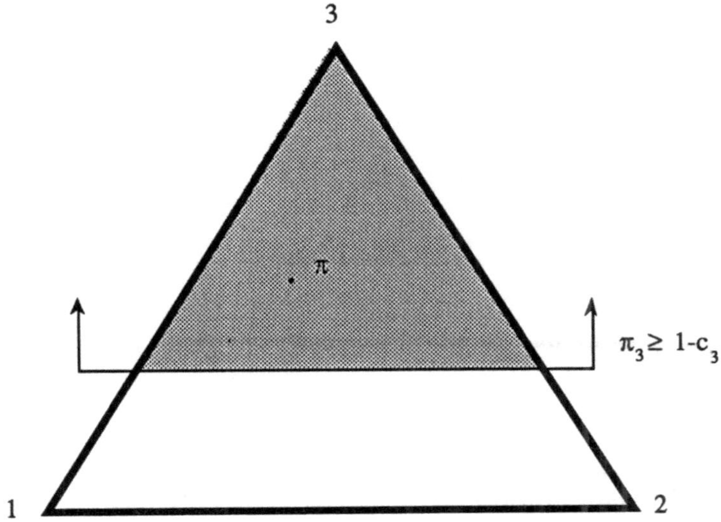

Figure 3.7 Feasibility condition for an imputation in an essential, nonconstant-sum game: $\pi_1+\pi_2 \leq v(\{1,2\})=c_3$. *This is equivalent to* $\pi_3 \geq 1-c_3$.

Through parallel reasoning, the feasibility criterion for domination through coalitions {1, 3} and {2, 3} require similar constraints. These are

$$\pi_2 \geq 1 - c_2$$
$$\pi_1 \geq 1 - c_1$$

where c_i denotes the characteristic value of the alliance omitting agent i; namely, $\{1, 2, 3\} - \{i\}$.

An imputation, to be feasible for domination by coalition $\mathbf{K} = \mathbf{N} - \{i\}$, needs to satisfy only the constraint

$$\pi_i \geq 1 - c_i$$

But it might also satisfy one or more of the remaining feasibility conditions. The imputation π in Figure 3.8 happens to satisfy all three feasibility conditions.

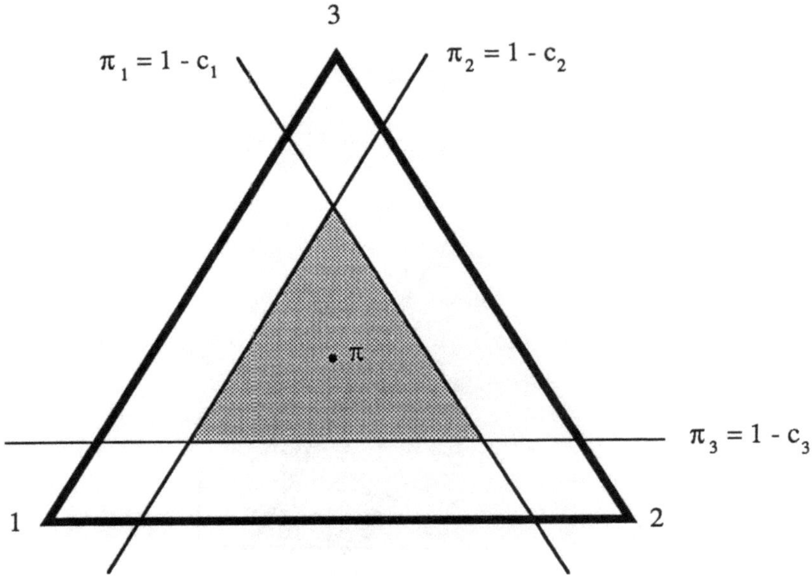

Figure 3.8 Shaded triangle represents imputations which satisfy the feasibility condition of domination by all three pairwise coalitions {1,2}, {1,3}, and {2,3}.

In addition to feasibility, the preference conditions must also be satisfied for dominance to hold. The allowed regions for the preference criterion were a set of parallelograms, as depicted in Figure 3.6 for constant-sum games.

Figure 3.9 depicts the set of dominated points which satisfy all the conditions of feasibility and preference. More specifically, feasibility for each coalition $\mathbf{N} - \{i\}$ requires

$$\pi_i \geq 1 - c_i$$

for $i \in \{1, 2, 3\}$. Further, preference implies $\pi_j \geq \rho_j$ for $j \in \mathbf{N} - \{i\}$. A point in the shaded area represents a feasible imputation dominated by π.

Core of a Game

Domination refers to the preference of an imputation over another by some coalition. In particular, suppose that imputation π dominates ρ. If one coalition advocates imputation ρ over π, the proposal is likely to be opposed by the alliance favoring the latter.

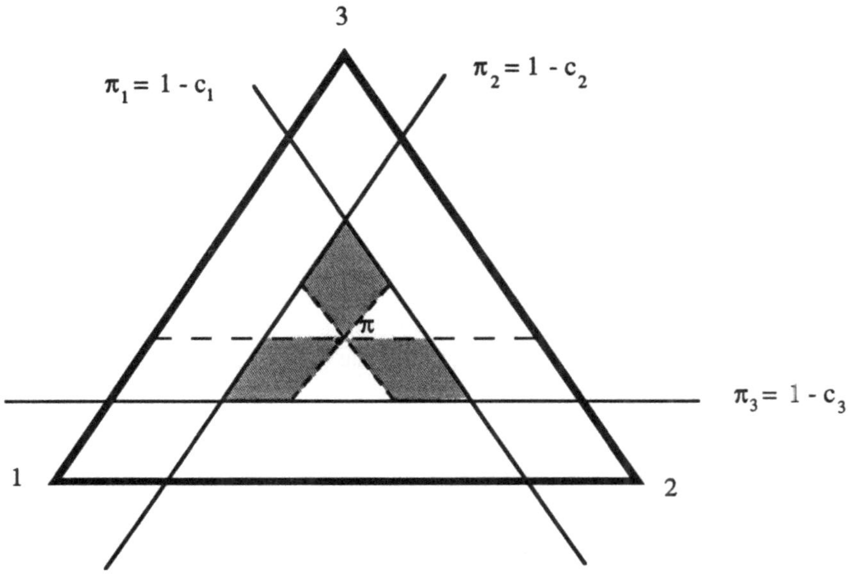

Figure 3.9 An imputation in the shaded region is a feasible point dominated by π.

An imputation which is not dominated by any other will not be opposed. Hence it is one of the more likely candidates to survive through the process of selecting an imputation for the population of agents. The set of nondominated imputations is called the *core* of a game.

One way to check whether an imputation lies in the core is to compare it against the characteristic function. If the collective payoff to each coalition matches or exceeds its characteristic value, then the imputation is undominated.

Theorem (Core Imputation). Let π be an imputation for a cooperative game $\langle \mathbf{N}, v \rangle$. Suppose **K** is any coalition from **N**. The imputation π belongs to the core iff the collective allocation for each **K** at least equals its characteristic value:

$$\sum_{i \in \mathbf{K}} \pi_i \geq v(\mathbf{K}) \qquad \text{(Eq. 17)}$$

Proof. Sufficiency is relatively straightforward to prove. Suppose that π were dominated by another imputation ρ. Then there is some coalition **K** for which

$$v(\mathbf{K}) \geq \sum_{i \in \mathbf{K}} \rho_i > \sum_{i \in \mathbf{K}} \pi_i$$

The first inequality above springs from the feasibility condition for ρ to be a dominating imputation. The second inequality implies that Eq. 17 is violated. This verifies the sufficiency part of the proof.

For necessity, we consider the game in its binary form. For these games, we know that $v(\mathbf{N}) = 1$ and $v(i) = 0$ for each $i \in \mathbf{N}$. Suppose that the relation

$$\sum_{i \in \mathbf{K}} \pi_i < v(\mathbf{K}) \qquad \text{(Eq. 18)}$$

holds for some coalition **K**. This coalition **K** cannot be a singleton, since individual rationality would be violated for π. At the other end of the spectrum, **K** cannot be the entire population since then group rationality would be violated.

By the complementarity property, we can write

$$\sum_{i \notin \mathbf{K}} \pi_i = v(\mathbf{N}) - \sum_{i \in \mathbf{K}} \pi_i \geq v(\mathbf{K}) - \sum_{i \in \mathbf{K}} \pi_i > 0$$

The first inequality above follows from the fact that $v(\mathbf{K})$ cannot exceed $v(\mathbf{N}) = 1$. The second inequality relies on Eq. 18.

Now consider a positive number ε for which

$$\varepsilon < \frac{1}{|\mathbf{K}|}\left[v(\mathbf{K}) - \sum_{i \in \mathbf{K}} \pi_i \right] \qquad \text{(Eq. 19)}$$

Here $|\bullet|$ denotes the cardinality of the enclosed set. Let $\overline{\mathbf{K}}$ denote the complement of K, namely $\overline{\mathbf{K}} \equiv \mathbf{N} - \mathbf{K}$. We can construct a new vector $\rho = (\rho_1, \ldots, \rho_n)$ in the following way:

$$\begin{array}{ll} \pi_i + \varepsilon & : i \in \mathbf{K} \\ \frac{1}{|\overline{\mathbf{K}}|}\left[\sum_{i \in \overline{\mathbf{K}}} \pi_i - |\mathbf{K}|\varepsilon \right] & : i \in \overline{\mathbf{K}} \end{array}$$

We can check that ρ is an imputation. For $i \in \mathbf{K}$,

$$\rho_i = \pi_i + \varepsilon > \pi_i \geq v(i)$$

where the last inequality follows from individual rationality. For $i \in \overline{\mathbf{K}}$, we can write

$$\begin{aligned}
\rho_i &= \frac{1}{|\overline{\mathbf{K}}|}\left[\sum_{i \in \overline{\mathbf{K}}} \pi_i - |\mathbf{K}|\,\varepsilon\right] \\
&> \frac{1}{|\overline{\mathbf{K}}|}\left[\sum_{i \in \overline{\mathbf{K}}} \pi_i - \left[v(\mathbf{K}) - \sum_{i \in \mathbf{K}} \pi_i\right]\right] \\
&= \frac{1}{|\overline{\mathbf{K}}|}\left[\sum_{i \in \mathbf{N}} \pi_i - v(\mathbf{K})\right] = \frac{1}{|\overline{\mathbf{K}}|}[v(\mathbf{N}) - v(\mathbf{K})] \geq 0
\end{aligned}$$

The first inequality follows from Eq. 19. The last inequality relies on the monotonic property of v. Since the game is in binary form, $v(i) = 0$. These results imply that $\rho_i \geq v(i)$, thereby satisfying individual rationality.

We can verify group rationality for ρ:

$$\begin{aligned}
\sum_{i \in \mathbf{N}} \rho_i &= \sum_{i \in \mathbf{K}} [\pi_i + \varepsilon] + \sum_{i \in |\overline{\mathbf{K}}|} \frac{1}{|\overline{\mathbf{K}}|}\left[\sum_{i \in |\overline{\mathbf{K}}|} \pi_i - |\mathbf{K}|\,\varepsilon\right] \\
&= \sum_{i \in \mathbf{K}} \pi_i + \sum_{i \in \overline{\mathbf{K}}} \pi_i = v(\mathbf{N})
\end{aligned}$$

Hence ρ is an imputation.

For each $i \in \mathbf{K}$, $\rho_i = \pi_i + \varepsilon > \pi_i$. Hence ρ dominates π by coalition K. Consequently, π would not belong to the core, a result which contradicts an initial assumption concerning π. This concludes the necessary part of the proof. ■

The preceding theorem implies that a necessary and sufficient condition for an imputation is that it satisfy a set of inequalities. The intersection of these inequalities, when it exists, is a convex region within the fundamental simplex. For the case of 3 agents, the convex region may be a 2-dimensional polygon, a line segment, or a point. However, it is also possible that the inequalities are mutually inconsistent, resulting in no solution. In other words, the core might be empty.

Examples of Cores

The simplest category of games to consider in connection with cores is that of inessential games. Every inessential game has a unique imputation. Since this imputation cannot be dominated, the core consists of this single point.

Essential games may be partitioned into constant-sum and nonconstant-sum types. The former type has no core, while the latter type exhibits a richer collection of possibilities.

We first prove the result for essential, constant-sum games.

Theorem. The core of an essential, constant-sum game is empty.

Proof. Suppose that an imputation π belongs to the core. Since π is an imputation, the game has at least 3 agents. From the Core Imputation Theorem, the following inequalities are satisfied:

$$\begin{aligned} \pi_i &\geq v(i) \\ \sum_{j \neq i} \pi_j &\geq v(i) + v(\mathbf{N} - i) \end{aligned} \qquad \text{(Eq. 20)}$$

for each agent i. By adding these inequalities we obtain

$$\sum_{i \in \mathbf{N}} \pi_i \geq v(i) + v(\mathbf{N} - i)$$

But the left side equals $v(\mathbf{N})$ due to group rationality. The right side also equals $v(\mathbf{N})$ since the game is constant-sum. The result is strict equality in the above relation:

$$\sum_{i \in \mathbf{N}} \pi_i = v(\mathbf{N})$$

This also implies strict equality in Eq. 20. In other words,

$$\pi_i = v(i)$$

for each agent. The last two equations imply that the characteristic function is additive. This in turn is tantamount to saying that the game is inessential. Thus a nonempty core cannot result from an essential game. ■

We now turn to the general 3-agent game for which the characteristic function may be constant-sum or not. As usual, we will focus on the binary form of the game:

$$v(1) = v(2) = v(3) = 0$$
$$v(\{1, 2, 3\}) = 0$$
$$v(\{1, 2\}) \equiv c_3$$
$$v(\{1, 3\}) \equiv c_2$$
$$v(\{2, 3\}) \equiv c_1$$

Here each c_i is a variable in the interval [0, 1].

From the Core Imputation Theorem, an imputation π belongs to the core if and only if these additional conditions apply:

$$\pi_1 + \pi_2 \geq c_3$$
$$\pi_1 + \pi_3 \geq c_2$$
$$\pi_2 + \pi_3 \geq c_1$$

Since the π_i sum to unity, the preceding conditions may be written as

$$\pi_i \leq 1 - c_i \qquad \text{(Eq.21)}$$

for $i \in \{1, 2, 3\}$. In terms of the fundamental triangle, the above inequality implies that the point π lies farther away from each vertex i than the boundary formed by $\pi_i^0 \equiv 1 - c_i$.

By summing Eq. 21 over all agents, we obtain

$$\pi_1 + \pi_2 + \pi_3 \leq 3 - c_1 - c_2 - c_3$$

Since the π_i add to 1, we can write

$$c_1 + c_2 + c_3 \leq 2 \qquad \text{(Eq. 22)}$$

This is a necessary condition for π to lie in the core. We can also show that it is a sufficient condition. Suppose that Eq. 22 is satisfied. Then we can find a set of values $\varepsilon_i \geq 0$ for which

$$\varepsilon_1 + \varepsilon_2 + \varepsilon_3 = 2 - (c_1 + c_2 + c_3)$$

We define the values $\pi_i \equiv 1 - c_i - \varepsilon_i$; these values satisfy Eq. 21. Hence Eq. 22 is a sufficient condition for $\pi = (\pi_1, \pi_2, \pi_3)$ to belong to the core.

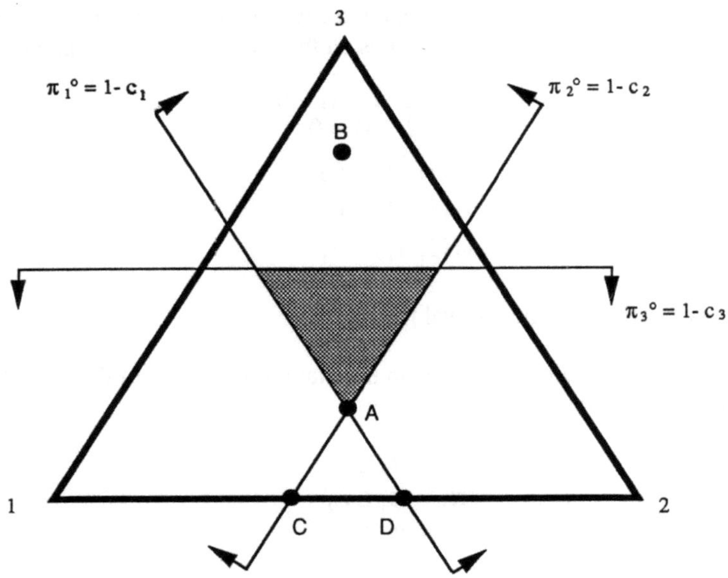

Figure 3.10 Example of a core for a general, 3-agent game. Since c_i lies in [0,1], so does π_1°. Hence the core can take the form of other polygons or even be empty.

In geometric terms, the core can take the form of a triangle as illustrated in Figure 3.10. Each characteristic value c_i can lie in the interval [0, 1]; as a result, so can the boundary line $\pi_i^0 \equiv 1 - c_i$. A consequence is that the core may be of dimension 2 or less. For instance, if the boundary line defined by π_3^0 intersects point *A* in the figure, then the core consists of a single point. If the line π_3^0 intersects point *B*, the result is a 5-sided polygon. Suppose that the lines π_1^0 and π_3^0 run through point *C*, while π_2^0 intersects point *D*; then the resulting core is the line segment *CD*.

Solution Principles

The imputations comprising the core of a game possess a measure of stability. In particular, there is no faction in the society of agents which would actively undermine efforts to adopt such an imputation.

On the other hand, lack of opposition does not imply advocacy. No faction may oppose a core imputation, but quite possibly none would champion it either. It would be desirable to have an active solution principle, one that identifies the best among a set of undominated imputations. Unfortunately, there is no such principle which seems generally applicable.

The best that is available is a collection of solution principles, each with its own merits as well as limitations. In the next few sections, we will examine several solution principles due to different writers.

Von Neumann-Morgenstern Solution. An imputation should satisfy a number of desirable attributes. One such attribute relates to stability: a candidate solution should not be precarious or readily subject to usurpation by another imputation.

The stability of a collection of solutions can be classified into internal and external facets. Internal stability refers to the lack of domination among imputations in a solution set. In contrast, external stability indicates that, for any imputation lying outside the solution set, a better imputation can be found within the set. These criteria represent the desiderata for solutions, as proposed by von Neumann and Morgenstern in their seminal monograph (von Neumann and Morgenstern, 1947).

To summarize, the *von Neumann-Morgenstern solution* to a cooperative game is the set **R** of imputations with the following attributes:

- *Internal Stability.* No imputation in the set **R** dominates any other imputation in the set.
- *External Stability.* Given any imputation lying outside **R**, an imputation within **R** can be found which dominates it.

The von Neumann-Morgenstern solution is also called the *VNM solution* or simply the *stable set.*

The VNM solution is a generalization of the concept of a core. More specifically, the core of a game, when it exists, lies within the VNM solution.

> **Theorem** (Core within VNM). Suppose that a cooperative game possesses both a core **C** and a VNM solution **R**. Then **C** is a subset of **R**.

Proof. Consider an imputation π lying in the core **C**. This imputation can dominate no other imputation in **C**; otherwise the latter should lie outside the core. Hence **C** is internally stable.

Suppose ρ is any imputation lying outside **C**. Then it must be dominated by some imputation $\pi \in C$; otherwise ρ should belong to the core. Consequently external stability is also assured, and **C** is a **VNM** solution. ■

We have seen in a previous section that an inessential game possesses exactly one imputation. This imputation represents both the core and the VNM solution.

What about an essential game? We recall that dominance cannot be imposed through a singleton or the entire population. Hence we can focus on a society of 3 or more agents.

As it happens, an essential, cooperative game cannot possess a VNM solution having precisely one imputation.

> **Theorem** (No VNM Singletons). Suppose that the VNM solution of a cooperative game consists of a single imputation. Then the game is inessential.

Proof. Suppose that π is an imputation for an essential game $\langle \mathbf{N}, v \rangle$. As is custom, we consider the binary form of this game. Since π is an imputation, it has some component
$\pi_i > 0$. By virtue of the preceding text paragraph, we can assume that the society contains 3 or more agents; that is, $n \equiv |\mathbf{N}| \geq 3$.

We define a new imputation $\rho = (\rho_1, \ldots, \rho_n)$ from the vector π by reallocating its ith component π_i evenly among all the other components. In algebraic terms:

$$\rho_i \equiv \begin{cases} 0 & :j = i \\ \pi_i + \dfrac{\pi_i}{n-1} & :j \neq i \end{cases}$$

Domination can only be supported by coalitions of two or more agents; since only the ith component of π exceeds that of ρ, π cannot dominate ρ.

Further, $\rho \neq \pi$. At this juncture we can consider two cases for imputation ρ:

1. It lies in the VNM solution.

2. It does not lie in the VNM solution. Since ρ is not dominated by π, it must be dominated by some other imputation σ in the VNM set.

Either way, the VNM solution consists of more than one imputation. Hence the game cannot be essential. ■

The principles behind the VNM solution represent an interesting attempt to characterize desirable imputations. On the other hand, the VNM solution has its limitations. Among these are the following:

- *Nonexistence.* Some cooperative games have no VNM solutions.

- *Multiplicity.* Many cooperative games possess VNM solutions containing more than one imputation. In fact, there is often an infinite number of candidates.

  In particular, by the preceding theorem, all essential cooperative games have at least two imputations in the VNM solution. The criterion of stability cannot further winnow this set to select an optimal imputation.

These attributes are perhaps better viewed as features rather than handicaps for the VNM solution concept. Many practical problems in engineering, management, economics, and governance often allow for multiple solutions or even none.

In the case of multiple solutions, additional criteria are required to distinguish the relative desirability among imputations. One such principle relates to fairness. We now turn to this topic.

Assumptions of the Shapley Solution

As we have seen, the core and the VNM solutions represent only partially successful methods for selecting among alternative imputations. These solution methods are both too lax and too strict: they may identify multiple solutions in certain cases, and none in others.

An alternative approach is the Shapley solution. This method is based on the general notion of fairness. The underlying assumptions pertain to payoff conservation, symmetry among the agents, and additivity of reward across games. This subsection presents the axioms of the Shapley approach while the next focuses on their implications.

Consider a cooperative game $\langle \mathbf{N},\nu\rangle$ consisting of n agents. The characteristic function ν fully defines the game for the society $\mathbf{N}=(1,2,...,n)$. Our objective is to identify a procedure φ_i for each agent which describes his allocation of the bounty resulting from the game. Since the game is defined by the characteristic function ν, we can write $\varphi_i=\varphi_i(\nu)$.

These procedures collectively define a vector $\varphi(\nu)$. That is,

$$\varphi(\nu)=\left(\varphi_i(\nu),\ldots,\varphi_n(\nu)\right)$$

Since φ is to represent a fair allocation of the benefits from the game, it should at least satisfy the conditions of individual and group rationality. In other words, the vector φ is an imputation.

We begin with some definitions on our journey to the first axiom of the Shapley approach. An agent i is called an *isolate* or a *dummy* if his participation has no effect on the total payoff to the other members of any coalition **K**. In other words, suppose **K** is any alliance which excludes agent i; then

$$\nu(\mathbf{K}\cup\{i\})=\nu(\mathbf{K})+\nu(i) \qquad \text{(Eq. 23)}$$

From this definition, we see that an isolate has no impact on the efficacy of a coalition, whether synergistically or detrimentally. What he offers the group is precisely what he can earn in isolation.

Suppose we consider the game in binary form. In this canonical format, the value of a singleton is nil; that is, $\nu(i) = 0$. Consequently, including or excluding any number of isolates has no impact on the strategic consequences of a cooperative game.

From a strategic perspective, then, the significant agents in a game are those who bring synergism. The set **S** which includes all nonisolates is called a *support* or *carrier* of the game.

If **S** denotes the support of the game, then the set **N** – **S** includes only isolates. For any coalition **K**, the characteristic value is that due to the supporting members in **K**, plus that of the isolates in the alliance:

$$v(\mathbf{K}) = v(\mathbf{S} \cap \mathbf{K}) + \sum_{i \in \mathbf{N}-\mathbf{S}} v(i)$$

We have seen that an isolate has no strategic impact on a coalition. As a result, a reasonable principle of fairness is for him to receive no payoff from the coalition, nor to contribute anything to the group. (This principle, reasonable as it appears, is of course at odds with certain practices in a welfare state. Among these are progressive taxation and income redistribution.)

We state the assumption of separating funds in the form of an axiom.

Axiom of Effectiveness. Let **S** be the support of the game $\langle \mathbf{N}, v \rangle$. Then the collective allocation for **S** matches its characteristic value:

$$\sum_{i \in \mathbf{S}} \varphi_i(v) = v(\mathbf{S}) \qquad \text{(Eq. 24)}$$

This conservation law stipulates that the collective allocation should neither exceed nor fall short of the community budget.

Consider the coalition $\mathbf{K} = \mathbf{S} \cup \{j\}$, where **S** is a support of the game and j an isolate not in **S**. Then $\mathbf{S} \cup \{j\}$ is also a support of the game. From Eq. 24 we can write

$$\sum_{i \in \mathbf{S} \cup \{j\}} \varphi_i(v) = v(\mathbf{S} \cup \{j\}) = v(\mathbf{S}) + v(j) = \sum_{i \in \mathbf{S}} \varphi_i(v) + v(j)$$

The second equality depends on Eq. 23, while the third relies on Eq. 24 once more. Comparing the first and last expressions leads to the result $\varphi_j(v) = v(j)$ for any isolate j. This deduction is consistent with the informal assumption that the allocation to an isolate should match his characteristic value.

A second principle of fairness relates to symmetry. Agents of equal ability should receive equal allocations. We introduce a couple of concepts before formalizing this assumption.

A *permutation* Π is a function which maps **K** onto itself. For each $i \in \mathbf{K}$, there is a unique object $\Pi(i)$ in the set. The codomain of the permutation is indicated by the notation $\Pi(\mathbf{K})$ or simply $\Pi\mathbf{K}$.

As an illustration, suppose $\mathbf{K}=\{1,2,3\}$. Then the function $\Pi(i) = i \bmod(3) + 1$ is a permutation of $\mathbf{K} = \{1, 2, 3\}$ into the set $\Pi\mathbf{K} = \{2,3,1\}$.

Consider a coalition **K** from the game $\langle \mathbf{N},v \rangle$ A permutation $\Pi\mathbf{K}$ is called an *automorphism* if the characteristic function is invariant under the mapping; more precisely,

$$v(\Pi\mathbf{K}) = v(\mathbf{K})$$

In other words, the order of the agents in a coalition has no impact on its efficacy.

We require that the function φ exhibit this property of invariance.

Axiom of Symmetry. Let Π be any automorphism of the game $\langle \mathbf{N}, v \rangle$. Then the allocation φ_i is invariant under the transformation:

$$\varphi_i(v) = \varphi_{\Pi(i)}(v)$$

This axiom stipulates a form of equality among agents.

We now consider two games $\langle \mathbf{N},v \rangle$ and $\langle \mathbf{N},v' \rangle$ involving the same society **N**. The practical interpretation is that of a society concurrently engaged in two separate games.

It seems plausible to define the independence of these games in terms of the additivity of their characteristic functions. That is, we will say that the games $\langle \mathbf{N},v \rangle$ and $\langle \mathbf{N},v' \rangle$ are *independent* if

$$(v + v')(\mathbf{K}) = v(\mathbf{K}) + v'(\mathbf{K})$$

for every coalition **K** in **N**.

A reasonable principle of fairness relates to the additivity of allocations for independent games.

Axiom of Additivity. Let $\langle \mathbf{N},v \rangle$ and $\langle \mathbf{N},v' \rangle$ be independent games. The allocation for the joint game is the sum of those due to each game:

$$\varphi(v + v') = \varphi(v) + \varphi(v')$$

This is also called the Axiom of Aggregation. It allows us to evaluate the allocations among multiple games.

An allocation φ which satisfies the Axioms of Effectiveness, Symmetry, and Additivity is called a ***Shapley vector*** or ***Shapley value***. The underlying axioms are both minimal and complete. For every cooperative game $\langle \mathbf{N}, v \rangle$, there is precisely one Shapley vector.

Properties of Shapley Vector

In this section, we explore some properties of the Shapley vector and a procedure for its construction. Consider a particular coalition **R** from a population **N**. The characteristic function $v_{\mathbf{R}}$ is called *simplest* if it serves as an indicator function for the inclusion of **R**. More specifically, $v_{\mathbf{R}}$ is defined for any coalition **K** from **N** in the following way:

$$v_{\mathbf{R}}(\mathbf{K}) = \begin{cases} 1 & :\mathbf{K} \supset \mathbf{R} \\ 0 & :\mathbf{K} \not\supset \mathbf{R} \end{cases}$$

In other words, $v_{\mathbf{R}}$ equals unity if **K** contains the critical coalition **R**, and equals zero otherwise.

We first consider Shapley vectors for any characteristic function of the form $cv_{\mathbf{R}}$, where c is a nonnegative constant and $v_{\mathbf{R}}$ is a simplest function.

From the definition of $v_{\mathbf{R}}$, we can see that the characteristic value for any agent i outside **R** is nil: $cv_{\mathbf{R}}(i) = 0$ for $i \notin \mathbf{R}$. Therefore **R** is the support for $cv_{\mathbf{R}}$.

If agent i is an isolate, we can write

$$\varphi_i(cv_{\mathbf{R}}) = cv_{\mathbf{R}}(i) = 0$$

Drawing on the Axiom of Effectiveness, we obtain

$$\sum_{i \in \mathbf{R}} \varphi_i(cv_{\mathbf{R}}) = cv_{\mathbf{R}}(\mathbf{R}) = c \qquad \text{(Eq. 25)}$$

The second equality relies on the fact that $v_{\mathbf{R}}$ is a simplest function.

The validity of the preceding equation is independent of the order of agents in **R**. Hence any permutation which involves the coalition **R** is an automorphism. From the Axiom of Symmetry, all the allocations $\varphi_i(cv_{\mathbf{R}})$ for $i \in \mathbf{R}$ are equal to each other; we denote this value by $\varphi_0(cv_{\mathbf{R}})$. We can sum over the agents in **R** to obtain

$$\sum_{i \in \mathbf{R}} \varphi_i(cv_{\mathbf{R}}) = |\mathbf{R}|\varphi_0(cv_{\mathbf{R}}) = cv_{\mathbf{R}}(\mathbf{R}) = c$$

The second equality relies on the Axiom of Symmetry, and the last on Eq. 25. Consequently, $\varphi_0(cv_R) = c/|\mathbf{R}|$.

To summarize, the Shapley vector for games of the form $cv_{\mathbf{R}}$ is defined by the components

$$\varphi_i(cv_{\mathbf{R}}) = \begin{cases} c/\mathbf{R} & : i \in \mathbf{R} \\ 0 & : \text{otherwise} \end{cases} \quad \text{(Eq. 26)}$$

Games in simplest form, as the nomenclature indicates, represent canonical structures. In fact, any cooperative game can be defined as a linear combination of these canonical forms.

Lemma (Linear Combination). Let $\langle \mathbf{N}, v \rangle$ be any cooperative game. The characteristic function v can be expressed as a linear combination of the simplest forms:

$$v = \sum_{\mathbf{R} \subset \mathbf{N}} \lambda_{\mathbf{R}} v_{\mathbf{R}} \quad \text{(Eq. 27)}$$

where each $\lambda_{\mathbf{R}}$ is a real number and $v_{\mathbf{R}}$ is a characteristic function in simplest form.

Proof. Suppose that Eq. 27 holds for some set of numbers $\lambda_{\mathbf{R}}$ and simplest functions $v_{\mathbf{R}}$. From the additivity of the characteristic functions, we can write

$$v(\mathbf{T}) = \sum_{\mathbf{R} \subset \mathbf{N}} \lambda_{\mathbf{R}} v_{\mathbf{R}}(\mathbf{T}) \quad \text{(Eq. 28)}$$

for any coalition **T** in **N**. Let **S** be a subset of **R**. Assume that **S** contains s agents, while **R** has r members. We define a new variable

$$\lambda_{\mathbf{R}} = \sum_{\mathbf{S} \subset \mathbf{R}} (-1)^{r-s} v(\mathbf{S}) \quad \text{(Eq.29)}$$

Inserting this expression for $\lambda_{\mathbf{R}}$ into Eq. 29 yields

$$v(\mathbf{T}) = \sum_{\mathbf{R} \subset \mathbf{N}} \left[\sum_{\mathbf{S} \subset \mathbf{R}} (-1)^{r-s} v(\mathbf{S}) \right] v_{\mathbf{R}}(\mathbf{T})$$

In this summation, all terms vanish except those for which **T** contains **R**; in formal terms, $\mathbf{R} \subset \mathbf{T}$. Moreover, since $v_{\mathbf{R}}$ is a simplest function, we have $v_{\mathbf{R}}(\mathbf{T}) = 1$ when $\mathbf{R} \subset \mathbf{T}$. Drawing on these arguments, we obtain

$$v(\mathbf{T}) = \sum_{\mathbf{R} \subset \mathbf{N}} \sum_{\mathbf{S} \subset \mathbf{R}} (-1)^{r-s} v(\mathbf{S})$$

Interchanging the summation yields

$$v(\mathbf{T}) = \sum_{\mathbf{S} \subset \mathbf{T}} \left[\sum_{\mathbf{S} \subset \mathbf{R} \subset \mathbf{T}} (-1)^{r-s} \right] v(\mathbf{S})$$

The first summation occurs over the coalitions **S**, and the second over **R**. Let t denote the cardinality of **T**. Then the expression in brackets can be written as a double summation by focusing on coalitions **R** having the same size:

$$\sum_{\mathbf{S} \subset \mathbf{R} \subset \mathbf{T}} (-1)^{r-s} = \sum_{r=s}^{t} \sum_{\substack{\mathbf{S} \subset \mathbf{R} \subset \mathbf{T} \\ |\mathbf{R}| = r}} (-1)^{r-s} = \sum_{r=s}^{t} \binom{t-s}{r-s} (-1)^{r-s}$$

The second equality follows from the selection of s members in a set of r objects from a population of size t. In particular, there are $\binom{t-s}{r-s}$ ways of choosing a set of size r which includes a specific set **S** out of a collection of **T** members.

We now insert the last result into the preceding expression for $v(\mathbf{T})$, and list separately the term for **S** = **T**. The result is

$$v(\mathbf{T}) = \sum_{\substack{\mathbf{S} \subset \mathbf{T} \\ \mathbf{S} \neq \mathbf{T}}} \left[\sum_{r=s}^{t} \binom{t-s}{r-s} (-1)^{r-s} \right] v(\mathbf{S}) + v(\mathbf{T})$$

The term within brackets is the binomial expansion of $(1-1)^{t-s}$, which equals zero. The expression on the right side is simply $v(\mathbf{T})$. Consequently, Eq. 28 is validated when $\lambda_{\mathbf{R}}$ is chosen as indicated by Eq. 29.

Having examined the existence of the canonical form, we now turn to its uniqueness. Suppose that two sets of coefficients $\lambda_{\mathbf{R}}$ and $\mu_{\mathbf{R}}$ exist when satisfy Eq. 28:

$$v(\mathbf{T}) = \sum_{\mathbf{R} \subset \mathbf{N}} \lambda_{\mathbf{R}} v_{\mathbf{R}}(\mathbf{T}) = \sum_{\mathbf{R} \subset \mathbf{N}} \mu_{\mathbf{R}} v_{\mathbf{R}}(\mathbf{T}) \qquad \text{(Eq. 30)}$$

We can show $\lambda_{\mathbf{R}}$ and $\mu_{\mathbf{R}}$ must match. The proof rests on induction on the supersets of **R**.

Consider the case $\mathbf{T} = \{i\}$ in Eq. 30. Then each term in the summations will vanish except for the situation $\mathbf{R} = \{i\}$. Since $v_\mathbf{R}$ is a simplest function, $v_{\{i\}}(i) = 1$, yielding $\lambda_\mathbf{R} = \mu_\mathbf{R}$.

For the inductive case, consider some coalition **S**. Assume that for each **R** contained strictly in **S**, the coefficients are equal; that is, $\lambda_\mathbf{R} = \mu_\mathbf{R}$ when $\mathbf{R}\subset\mathbf{S}$ and $\mathbf{R}\neq\mathbf{T}$.

We consider the situation $\mathbf{S} = \mathbf{T}$ in Eq. 30. We can expand each summation for the three cases where **R** lies strictly within **S**, $\mathbf{R} = \mathbf{S}$, and $\mathbf{R}\not\subset\mathbf{S}$:

$$\sum_{\substack{\mathbf{R}\subset\mathbf{S}\\ \mathbf{R}\neq\mathbf{S}}} \lambda_\mathbf{R} v_\mathbf{R}(\mathbf{S}) + \lambda_\mathbf{R} v_\mathbf{R}(\mathbf{S})\Big|_{\mathbf{R}=\mathbf{S}} + \sum_{\mathbf{R}\not\subset\mathbf{S}} \lambda_\mathbf{R} v_\mathbf{R}(\mathbf{S}) =$$

$$\sum_{\substack{\mathbf{R}\subset\mathbf{S}\\ \mathbf{R}\neq\mathbf{S}}} \mu_\mathbf{R} v_\mathbf{R}(\mathbf{S}) + \mu_\mathbf{R} v_\mathbf{R}(\mathbf{S})\Big|_{\mathbf{R}=\mathbf{S}} + \sum_{\mathbf{R}\not\subset\mathbf{S}} \mu_\mathbf{R} v_\mathbf{R}(\mathbf{S})$$

By the inductive hypothesis, $\lambda_\mathbf{R} = \mu_\mathbf{R}$ when **R** lies strictly in **S**. On the other hand, $v_\mathbf{R}$ is a simplest function; if $\mathbf{R}\not\subset\mathbf{S}$, then some critical agents are missing from **S** and $v_\mathbf{R}(\mathbf{S}) = 0$. These conditions imply that $\lambda_\mathbf{R} = \mu_\mathbf{R}$ when $\mathbf{R} = \mathbf{S}$.

Since this result is true for any set **S**, we see that $\lambda_\mathbf{R} = \mu_\mathbf{R}$ for each **R**. Hence the representation $v = \Sigma_{\mathbf{R}\subset\mathbf{N}}\lambda_\mathbf{R} v_\mathbf{R}$ is unique. ■

Nothing has been said thus far of the sign of any coefficient $\lambda_\mathbf{R}$ in Eq. 28. As it happens, we may speak of negative coefficients in the linear combination. The validity of such practice is demonstrated in the next lemma.

> **Lemma** (Difference of Characteristic Functions). Suppose that v, v' and $v - v'$ are characteristic functions. Then the Shapley value is linear in the difference of functions:

$$\varphi(v - v') = \varphi(v) - \varphi(v')$$

Proof. We can write $v = [v - v'] + v'$. From the Axiom of Additivity,

$$\varphi(v) = \varphi([v - v'] + v') = \varphi(v - v') + \varphi(v')$$

The desired result follows immediately. ■

Determination of Shapley Vector. We now address the computation of the Shapley vector for an arbitrary characteristic function. From the Linear Combination Lemma, any characteristic function v can be written as

$$v = \sum_{\mathbf{R} \subset \mathbf{N}} \lambda_{\mathbf{R}} v_{\mathbf{R}}$$

where each $v_{\mathbf{R}}$ is a simplest function.

We first segregate the summation in terms of semipositive and negative coefficients

$$v = \sum \lambda_{\mathbf{R}} v_{\mathbf{R}} \Big|_{\lambda_{\mathbf{R}} \geq 0} - \sum (-\lambda_{\mathbf{R}}) v_{\mathbf{R}} \Big|_{\lambda_{\mathbf{R}} < 0} \qquad \text{(Eq. 31)}$$

It is understood that the summation occurs for all coalitions $\mathbf{R}$ in $\mathbf{N}$. If $v_{\mathbf{R}}$ is a characteristic function and $\lambda_{\mathbf{R}}$ a positive constant, so is $\lambda_{\mathbf{R}} v_{\mathbf{R}}$; in fact, we recall from an earlier section that $\lambda_{\mathbf{R}} v_{\mathbf{R}}$is strategically equivalent to $v_{\mathbf{R}}$. The sum $\Sigma \lambda_{\mathbf{R}} v_{\mathbf{R}}$ of characteristic functions is a characteristic function as well. Finally, from the preceding lemma, the difference of characteristic functions given on the right side of Eq. 31 is also a characteristic function. We can therefore write

$$\varphi_i(v) = \varphi_i \left[\sum_{\mathbf{R} \subset \mathbf{N}} \lambda_{\mathbf{R}} v_{\mathbf{R}} \Big|_{\lambda_{\mathbf{R}} \geq 0} \right] - \left[\sum_{\mathbf{R} \subset \mathbf{N}} (-\lambda_{\mathbf{R}}) v_{\mathbf{R}} \Big|_{\lambda_{\mathbf{R}} < 0} \right]$$

$$= \sum_{\mathbf{R} \subset \mathbf{N}} \varphi_i \left(\lambda_{\mathbf{R}} v_{\mathbf{R}} \Big|_{\lambda_{\mathbf{R}} \geq 0} \right) - \sum_{\mathbf{R} \subset \mathbf{N}} \varphi_i \left(-\lambda_{\mathbf{R}} v_{\mathbf{R}} \Big|_{\lambda_{\mathbf{R}} < 0} \right)$$

The second equality relies on the Axiom of Linearity. The use of Eq. 26 leads to

$$\varphi_i(v) = \sum_{i \in \mathbf{R} \subset \mathbf{N}} \frac{\lambda_{\mathbf{R}}}{|\mathbf{R}|} \Bigg|_{\lambda_{\mathbf{R}} \geq 0} - \sum_{i \in \mathbf{R} \subset \mathbf{N}} \frac{-\lambda_{\mathbf{R}}}{|\mathbf{R}|} \Bigg|_{\lambda_{\mathbf{R}} < 0}$$

$$= \sum_{i \in \mathbf{R} \subset \mathbf{N}} \frac{\lambda_{\mathbf{R}}}{|\mathbf{R}|}$$

Inserting the expression for $\lambda_{\mathbf{R}}$ from Eq. 29 yields

$$\varphi_i(v) = \sum_{i \in \mathbf{R} \subset \mathbf{N}} \frac{1}{|\mathbf{R}|}\left[\sum_{\mathbf{S} \subset \mathbf{R}} (-1)^{r-s} v(\mathbf{S})\right]$$
$$= \sum_{\mathbf{S} \subset \mathbf{N}} \left[\sum_{\substack{\mathbf{R} \subset \mathbf{N} \\ \mathbf{R} \supset (\mathbf{S} \cup i)}} \frac{1}{|\mathbf{R}|} (-1)^{r-s}\right] v(\mathbf{S}) \quad \text{(Eq.32)}$$

The second equality follows from an exchange of summation operations. The inner sum within brackets is fixed for a particular agent i and coalition $\mathbf{S}$; let $\alpha_i(\mathbf{S})$ denote this sum:

$$\alpha_i(\mathbf{S}) \equiv \sum_{\substack{\mathbf{R} \subset \mathbf{N} \\ \mathbf{R} \supset (\mathbf{S} \cup i)}} \frac{1}{|\mathbf{R}|} (-1)^{r-s} \quad \text{(Eq. 33)}$$

Suppose that i is a member of $\mathbf{S}$. Then we can write

$$\alpha_i(\mathbf{S} - i) = \sum_{\substack{\mathbf{R} \subset \mathbf{N} \\ \mathbf{R} \subset (\mathbf{S} - i) \cup i}} \frac{1}{|\mathbf{R}|} (-1)^{r - |\mathbf{S} - i|} = \sum_{\substack{\mathbf{R} \subset \mathbf{N} \\ \mathbf{R} \subset \mathbf{S}}} \frac{1}{|\mathbf{R}|} (-1)^{r-s+1}$$
$$= - \sum_{\substack{\mathbf{R} \subset \mathbf{N} \\ \mathbf{R} \subset \mathbf{S}}} \frac{1}{|\mathbf{R}|} (-1)^{r-s} = -\alpha_i(\mathbf{S}) \quad \text{(Eq. 34)}$$

The outer summation in Eq. 32 occurs for each $\mathbf{S}$ in $\mathbf{N}$. We segregate the summation by pairs, in terms of the coalition $\mathbf{S}$ which contains agent i, in conjunction with $\mathbf{S} - \{i\}$. This yields

$$\varphi_i(v) = \sum_{\substack{i \in \mathbf{S} \\ \mathbf{S} \subset \mathbf{N}}} [\alpha_i(\mathbf{S}) v(\mathbf{S}) + \alpha_i(\mathbf{S} - i) v(\mathbf{S} - i)]$$
$$= \sum_{\substack{i \in \mathbf{S} \\ \mathbf{S} \subset \mathbf{N}}} \alpha_i(\mathbf{S}) [v(\mathbf{S}) - v(\mathbf{S} - i)] \quad \text{(Eq. 35)}$$

The last equality depends on Eq. 34. Since $\mathbf{S}$ contains i, we can write Eq. 33 as

$$\varphi_i(\mathbf{S}) = \sum_{\substack{\mathbf{R}\subset \mathbf{N}\\ \mathbf{R}\supset \mathbf{S}}} \frac{1}{|\mathbf{R}|}(-1)^{r-s} = \sum_{r=s}^{n} \frac{1}{r} \sum_{\substack{\mathbf{R}\subset \mathbf{N}\\ \mathbf{R}\supset \mathbf{S}\\ |\mathbf{R}|=r}} (-1)^{r-s}$$

For the second equality, we have split up the original summation in terms of coalitions **R** of fixed size r. The last summation involves a constant term $(-1)^{r-s}$, of which there are $\binom{n-s}{r-s}$ replications. We can therefore write

$$\alpha_i(\mathbf{S}) = \sum_{r=s}^{n} \frac{1}{r}\binom{n-s}{r-s}(-1)^{r-s}$$

The evaluation of this summation is a short exercise in calculus. Since $r > 0$, we can replace the factor 1/r by its integral equivalent in the form of $\int_0^1 x^{r-1}\mathrm{d}x$. The result is

$$\begin{aligned}\alpha_i(\mathbf{S}) &= \sum_{r=s}^{n}\left[\int_0^1 x^{r-1}\mathrm{d}x\right]\binom{n-s}{r-s}(-1)^{r-s}\\ &= \int_0^1\left\{\sum_{r=s}^{n} x^{r-s}\binom{n-s}{r-s}(-1)^{r-s}\right\}x^{s-1}\mathrm{d}x \qquad \text{(Eq. 36)}\\ &= \int_0^1\left\{1-x^{n-s}\right\}x^{s-1}\mathrm{d}x\end{aligned}$$

For the second equality, the order of integration has been moved and the factor x^{r-1} replaced by $x^{r-s}x^{s-1}$. The third equality relies on the binomial formula for expanding two terms raised to an exponent.

Let us denote the last expression, called a Beta function, as B(n, s). We first note that

$$B(n,1) \equiv \int_0^1 (1-x)^{n-1}\mathrm{d}x = -\left.\frac{(1-x)^n}{n}\right|_0^1 = \frac{1}{n} \qquad \text{(Eq. 37)}$$

By defining new variables $u \equiv (1-x)^{n-s}$ and $v = x^s / s$, we can integrate $B(n, s)$ by parts:

$$B(n,s) = \int_0^1 (1 - x)^{n-s} d\left(\frac{x^s}{S}\right)$$

$$= (1 - x)^{n-s}\left(\frac{x^s}{S}\right)\Bigg|_0^1 + \int_0^1 \left(\frac{x^s}{S}\right)(n - s)(1 - x)^{n-s-1}dx$$

$$= \frac{n - s}{s} B(n, s+1)$$

We obtain a recursive formula

$$B(n,s+1) = \frac{s}{n - s} B(n,s)$$

Applying this to Eq. 37 leads to the expression

$$B(n, 2) = \frac{1}{n - 1} - \frac{1}{n}$$

More generally, we have $B(n,s) = (n-s)!/n!$. From Eq. 36 we obtain

$$\alpha_i(\mathbf{S}) = \frac{(n - s)!\,(s - 1)!}{n!}$$

A return to Eq. 35 yields

$$\varphi_i(v) = \sum_{\mathbf{S} \ni i} \frac{(n-s)!\,(s-1)!}{n!} \left[v(\mathbf{S}) - v(\mathbf{S} - i)\right] \tag{Eq. 39}$$

Our remaining task is to verify that the allocation vector φ satisfies the axioms which were used for its derivation. To this end, we first demonstrate that individual rationality holds for $\varphi(v)$.

We recall that the characteristic function v is superadditive. Consequently, for any agent i in $\mathbf{S}$:

$$v(\mathbf{S}) - v(\mathbf{S} - i) \geq v(i) \tag{Eq. 39}$$

Using this inequality, we can write Eq. 38 as

$$\varphi_i(v) \geq \sum_{i \in \mathbf{S} \subset \mathbf{N}} \frac{(n\text{-}s)!\,(s\text{-}1)!}{n!} v(i)$$
$$= v(i) \sum_{s=1}^{n} \sum_{\substack{i \in \mathbf{S} \subset \mathbf{N} \\ |\mathbf{S}| = s}} \frac{(\text{n-s})!\,(\text{s-}1)!}{\text{n}!}$$

The equality follows from moving the fixed factor $v(i)$ to the front, and also by partitioning the summation in terms of sets of **S** having fixed size s. The term within the second summation is fixed for a given value of s. Further, there are $\binom{n-1}{s-1}$ ways of choosing a set of size n, which must contain a particular agent i, out of a society of n members. The preceding inequality becomes

$$\varphi_i(v) \geq v(i) \sum_{s=1}^{n} \left[\frac{(n-1)!}{(s-1)!\,(n-s)!}\right] \frac{(n-s)!\,(s-1)!}{n!}$$
$$= v(i) \sum_{s=1}^{n} \frac{1}{n} = v(i) \qquad \text{(Eq. 40)}$$

This verifies individual rationality for the allocation vector $\varphi(v)$.

The Axiom of Effectiveness can now be verified as follows. Suppose that agent i is an isolate. This implies

$$v(i) = v(\mathbf{S}) - v(\mathbf{S} - i)$$

for any coalition **S**. Hence Eq. 39 holds with strict equality. The succeeding inequalities also follow with strict equality, culminating in Eq. 40. Consequently,

$$\varphi_i(v) = v(i)$$

Let **S** be the support of **N**, and **N** – **S** the collection of isolates. Then

$$v(\mathbf{N}) = v(\mathbf{S}) + \sum_{i \in \mathbf{N}-\mathbf{S}} v(i) = \sum_{i \in \mathbf{S}} \varphi_i(v) + \sum_{i \in \mathbf{N}-\mathbf{S}} \varphi_i(v)$$

Since $v(i) = \varphi_i(v)$ for each $i \in \mathbf{N} - \mathbf{S}$, we obtain $v(\mathbf{S}) = \Sigma_{i \in \mathbf{S}}\, \varphi_i(v)$, thereby verifying the Axiom of Effectiveness.

For the Axiom of Symmetry, let Π be an automorphism of v_1. For any **S** in **N**, $v(\Pi\mathbf{S}) = v(\mathbf{S})$. When **S** successively denotes every coalition in **N**, so does $\Pi\mathbf{S}$. Applying Eq. 38 to the permutation $\Pi\mathbf{S}$, we obtain

$$\varphi_{\Pi(i)}(v) = \sum_{\Pi S \ni \Pi(i)} \frac{(n - s)!\,(s - 1)!}{n!} [v(\Pi \mathbf{S}) - v(\Pi(\mathbf{S} - i))]$$

We have made use of the fact that $|\Pi \mathbf{S}| = |\mathbf{S}| = s$. Since the characteristic function takes no account of the ordering among agents, $v(\Pi \mathbf{S}) = v(\mathbf{S})$.

Further, the coalitions $\Pi\mathbf{S}$ which include $\Pi(i)$ have a 1:1 correspondence with the coalitions **S** which encompass agent i. Hence the index of summation can be changed to $\mathbf{S} \ni \{i\}$. The result is

$$\varphi_{\Pi(i)}(v) = \sum_{\mathbf{S} \ni \{i\}} \frac{(n-s)!\,(s-1)!}{n!} [v(\mathbf{S}) - v(\mathbf{S} - \{i\})] = \varphi_i(v)$$

which verifies the property of symmetry.

To confirm the Axiom of Linearity, we recall that $\varphi_i(v)$ can be written as a linear combination of characteristic functions, as illustrated by the last equation. But characteristic functions can themselves combine linearly. These facts imply the linearity of the allocation vector $\varphi(v)$.

The discussion thus far demonstrates that Shapley vectors exist and are unique for each cooperative game $\langle \mathbf{N}, v\rangle$. Moreover, the Shapley vector can be constructed from the characteristic function by the use of Eq. 38.

The Shapley vector can be interpreted in terms of expected payoffs. This interpretation also allows for an alternative method for constructing the vector.

To this end, consider some permutation Π on the population **N**. We may denote the permuted list as $(\Pi(1),..., \Pi(n))$. Let $P[\Pi(i)]$ denote the *predecessor list* of $\Pi(i)$; that is

$$P[\Pi(i)] \equiv [\Pi(1), \Pi(2),..., \Pi(i-1)]$$

The characteristic value $v(P[\Pi(i)])$ is the collective payoff to the predecessors of agent i. In a similar way, $v(P[\Pi(i)] \cup \{i\})$ is the collective payoff when agent i joins his predecessors. The *incremental payoff* or *admission value* for the agent is given by

$$\Delta v(\Pi, i) \equiv v\left(P[\Pi(i)] \cup \{i\}\right) - v\left(P[\Pi(i)]\right) \qquad \text{(Eq. 41)}$$

This quantity represents the contribution attributable to agent i when he joins the group under permutation Π.

Suppose that all permutations are equally likely. Then the average value of the incremental contribution for agent i is precisely his Shapley value.

Theorem (Expected Contribution). Suppose that all permutations Π have the same probability of occurrence among the agents in a cooperative game $\langle N, v \rangle$. Then the expected value of the incremental contribution for each agent matches his Shapley value:

$$E_{\Pi}[\Delta v(\Pi,i)] = \frac{1}{n!}\sum_{n} \Delta v(\Pi,i) = \varphi_i(v)$$

Here E_{Π} indicates the expectation operator over the permutations ΠN.

Proof. Let **S** be any subset of population **N**. There are $n!$ permutations of the agents in **N**. Among these permutations, how many exhibit the sequence $\langle S, i, N-S \rangle$; that is, the agents in **S**, followed by i, followed by the agents outside **S**? There are

$$s!\ (1!)\ (n-s-1)!$$

such permutations, where s denotes the cardinality of **S**. The probability of observing such a permutation is given by the ratio of the preceding numbers:

$$P\left(\left\langle S, i, N - S \right\rangle\right) = \frac{s!\ (n-s-1)!}{n!}$$

For the sequence $\langle S, i, N-S \rangle$, the incremental payoff defined by Eq. 41 can be written as

$$\Delta v(\Pi, i) = v(S \cup i) - v(S)$$

Applying the expectation operator over the range of coalitions **S** yields

$$E_{\Pi}\Delta v(\Pi,i) = \sum_{S \subset (N - \{i\})} \frac{s!\ (n - s - 1)!}{n!}\left[v(S \cup \{i\}) - v(S)\right]$$

Being the index of summation, the coalition **S** is a kind of dummy variable in the preceding equation. We can change variables by letting **S** replace $S \cup \{i\}$. The equation becomes

$$E_{\Pi}\Delta v(\Pi,i) = \sum_{S \subset N} \frac{(s - 1)!\ (n - s)!}{n!}\left[v(S) - v(S - \{i\})\right] = \varphi_i(v)$$

Hence the Shapley value for agent i is simply the expected value of his incremental contribution to a coalition. ■

Another way to view the Shapley vector is as follows. Suppose that the agents in a population arrive in random order at a meeting spot. The admission value $\Delta v(\Pi, i)$ is the marginal contribution of agent i to the group which has already assembled before his arrival. The admission value depends not only on the identity of agent i, but on the order of arrivals–that is, the particular permutation Π. If all permutations are equally likely, then the expected admission value $E_{\Pi}\Delta v(\Pi, i)$ for agent i is precisely his Shapley value.

Application to Supervisor and Workers

A materials handling system consists of n agents. One of these is the supervisory controller, and the others are shuttles of identical capability.

Suppose that the shuttles by themselves lack the knowledge to discharge their role. On the other hand, the supervisor by itself has the knowledge but no physical capacity to move workpieces. A functional team consists of the supervisor plus at least one shuttle; in this case, the collective payoff $f(k)$ depends on the number k of shuttles in the coalition.

Let $1, 2, ..., n-1$ denote the shuttles and n the supervisor. Then the characteristic function can be written as

$$v(\mathbf{K}) = \begin{cases} f(k-1) & : n \in k \\ 0 & : \text{otherwise} \end{cases}$$

where k denotes the number of agents in $\mathbf{K}$.

Suppose that all permutations are equally likely. Then the supervisor will show up at position k, for any $k \in \{1, ..., n\}$, with probability $1/n$. Since there will be $k-1$ shuttles in the predecessor list, the admission value of the supervisor will be $f(k-1)$. The expected contribution can be obtained by considering all possible positions for the supervisor:

$$\varphi_n(v) = \frac{1}{n}\sum_{k=1}^{n} f(k\text{-}1) = \frac{1}{n}\sum_{k=1}^{n\text{-}1} f(k)$$

The second equality relies on a simple change of variables. Further, we make use of the fact that $f(0)$ must equal 0, from the preceding definition of $v(\mathbf{K})$.

The characteristic value for the entire population is $v(\mathbf{N}) = f(n-1)$. By the Axiom of Effectiveness, the collective allocation for all the shuttles is

$$\sum_{i=1}^{n\text{-}1} \varphi_i(v) = f(n\text{-}1) - \frac{1}{n}\sum_{k=1}^{n\text{-}1} f(k)$$

The Axiom of Symmetry implies that $\varphi_i(v) = \varphi_j(v)$ for all $i \neq j \neq n$. Consequently the Shapley value for each shuttle is

$$\varphi_i(v) = \frac{1}{n-1}\left[f(n-1) - \frac{1}{n}\sum_{k=1}^{n-1} f(k)\right]$$

As a special case, suppose that the payoff to any functional team is directly proportional to the number of shuttles. Then we can write $f(k) = ck$ for some positive constant c. Then the Shapley value for the supervisor is

$$\varphi_n(v) = \frac{1}{n}\sum_{k=1}^{n-1} ck = \frac{c(n-1)}{2}$$

The expected contribution for each shuttle $i \in \{1, 2, \ldots, n-1\}$ is

$$\varphi_i(v) = \frac{1}{n-1}\left[c(n-1) - \frac{c(n-1)}{2}\right] = \frac{c}{2}$$

The expected contribution of any shuttle is only half of its actual amount c when it does have an effect.

The ratio of Shapley values

$$\frac{\varphi_n(v)}{\varphi_i(v)} = n-1$$

gives a measure of the relative power of the supervisor. This metric is directly proportional to the population of shuttles. For the materials handling system described here, a fair partition of the overall payoff requires a disproportionate allocation to the supervisor, a disparity which increases with the number of agents under its control.

Learning systems and nested agents

The use of economic factors for coordinating multiple agents is compatible with the notion of a learning system whose performance improves over time. This property is due to the survival of the fittest in the economic arena.

If the performance of an agent is measured by the difference between payoff and cost, then an ineffectual agent is one that goes bankrupt. What happens to such an agent? In the economic realm, a corporation goes bankrupt and may be dissolved; in the biological sphere, an animal falls prey to predation, starvation, or disease. In an engineered domain, a bankrupt agent may be decommissioned – or equivalently, the hardware retained and the software replaced by a copy from a "fitter" agent.

Through the process of environmental selection, the effectiveness of the overall system can improve over time. This global learning process can occur even if the performance of any particular agent remains constant over time.

On the other hand, learning can also occur at the level of the individual agent. This may be achieved by strengthening the procedures which contribute to effective behavior, and weakening those that do not (Fu, 1968; Holland, 1975, 1986; Holland and Reitman, 1978; Kohonen, 1984; Kompass and Williams, 1983; Narendra and Thathachar, 1974; Shapiro and Narendra, 1969; Tyspkin, 1971; Valiant, 1984; Young and Fu, 1986). This approach is tantamount to the consideration of each procedure or morsel of knowledge as a subagent. A set of alternate procedures for accomplishing a particular task can be evaluated explicitly through financial factors. The income to a subagent will depend on the payoff for the overall agent, while the cost may be attributed to the use of resources such as processor time, memory space, energy consumption, or other factors.

It is clear that the explicit valuation of agents can be carried both up and down the hierarchy of decision making entities. In the downward direction, an agent may comprise a subagent of a higher-level object, which in turn comprises part of a larger system. In the other direction, a system such as a mobile robot may be regarded as an agent within a materials handling system, which in turn may be viewed as an agent of a manufacturing system.

In this way, the notion of an agent as a decision making entity is a versatile concept. One reflection of this versatility is the way in which the tools and techniques of learning systems can be applied to particular agents as well as their components or aggregates.

Conclusion and discussion

The limitations of centralized control for coordinating complex systems of autonomous agents, underscores the need for decentralized strategies. If the performance of an agent is to be evaluated thoroughly, its contribution to global objectives must be ascertained as well as its husbandry of limited resources. A performance measure based on an explicit valuation of benefits and costs leads to an economic perspective of system behavior.

The economic viewpoint allows us to draw on models and techniques from disparate fields such as game theory and political decision making. Some critical results from neighboring fields were examined, and their implications explored in the context of multiagent systems ranging from concurrent design to robotic control.

Other applications include the use of game theory for planning in production environments (Nof, 1991; Rajan and Nof, 1990). A set of manufacturing cells working toward a common goal can be modeled as a cooperative game. On the other hand, agents may need to compete for bottleneck resources such as expensive shuttles. These types of situations are better described in terms of competitive games.

The explicit valuation of action and the use of collaborative models facilitates the implementation of learning capability. This perspective is the subject of the next chapter.

An agent which performs well, acquires more capital and can control more resources. Conversely, an incompetent agent becomes bankrupt and eventually exits the competitive arena– at least in terms of software, even if the hardware is recycled.

A topic for future work pertains to the harshness or forgiving nature of the competitive arena. Should a poor performer be eliminated immediately, or be given a second chance? Should a handful of currently ineffectual agents be supported by the population, despite their carrying costs, in the interest of resiliency against future environmental changes which may favor their talents? If so, how many agents should be favored, and which ones?

Along these lines, some lessons can be drawn from the social and life sciences such as anthropology and evolutionary biology. When an environmental shift occurs, the extant species can follow three major paths: become extinct, adapt to the change, or metamorphose into a new species.

An example can be found in our own ancestry. About four million years ago, the jungles of Africa were populated by hominoids whom we now call australopithecines. These hominoids were more than ape but less than human by criteria such as bone structure and brain capacity. In the climatic changes that ensued, the jungles gave way to savannah, forcing the hominoids from the trees to the plains. There was little room for the australopithecines to persist as they were. As a result, they seem to have bifurcated into two new branches in the evolutionary tree.

One of these, *australopithecus boisei*, had a brain capacity larger than modern humans. But their bone structure was heavier and more ape-like than their direct ancestors in many other ways. As far as we can tell, this species met with extinction.

The second branch led to the first hominids about two million years ago, eventually resulting in the *Homo sapien.* This species continues to persist, and even to thrive.

Why did the original australopithecines bifurcate? Why did they diverge rather than evolve as a single species? Had the robust *boisei* survived, would they now be better equipped than humans to deal with future environmental shifts? At this stage of our comprehension, the answers still elude us.

It is interesting to note, however, that work from such disparate fields as biology, anthropology, economics, and political science can be brought to bear on the design of synthetic systems. These neighboring disciplines can provide motivating questions and at least suggestive propositions to guide the design of multiagent systems. And perhaps, in the years ahead, work in artificial systems will contribute in turn to the advancement of the natural and social sciences.

References

Arrow, K. "Control in Large Organizations." *Management Science*, v. 10(3), 1964: 397-408.

Arrow, K.J. *Social Choice and Individual Values.* NY: Wiley, 1951.

Arrow, K.J. and L. Hurwicz. "An Optimality Criterion for Decision Making under Ignorance." In K.J. Arrow and L. Hurwicz, eds., *Studies in Resource Allocation Processes*. London: Cambridge Univ. Press, 1977, pp. 463-471.

Arrow, K.J. and R. Radner. "Allocation of Resources in Large Teams." *Econometrica*, v. 47(2), 1979: 361-385.

Aubin, J.-P. *Mathematical Methods of Game and Economic Theory.* NY: North-Holland, 1979.

Boettcher, K.L. and A.H. Levis. "Modeling the Interacting Decisionmaker with Bounded Rationality." *IEEE Trans. on Systems, Man and Cybernetics*, v. SMC-12(3), May 1982: pp. 334-344.

Chang, T.S. and Y.-C. Ho. "Stochastic Stackelberg Games: Nonnested Multistage, Multiagent Incentive Problems." *IEEE Trans. Automatic Control*, v. AC-28(4), 1983: 477-488.

Chow, G.C. *Analysis and Control of Dynamic Economic Systems.* NY: Wiley, 1975.

Condon, Anne. *Computational Models of Games.* Cambridge, MA: MIT Press, 1989.

DeGroot, M.H. *Optimal Statistical Decisions.* New York: McGraw-Hill, 1970.

Farley, Arthur M. "A Probabilistic Model for Uncertain Problem Solving." *IEEE Transactions on Systems, Man, and Cybernetics*, v. 13(4), 1983, pp. 568-579.

Feldman, A. *Welfare Economics and Social Choice Theory.* Boston: Martinus Nijhof, 1980.

Ferguson, T.S. *Mathematical Statistics: A Decision Theoretic Approach.* NY: Academic, 1967.

Friedman, J. W. *Game Theory with Applications to Economics.* NY: Oxford U. Press, 1986.

Fu, K.S. *Sequential Methods in Pattern Recognition and Machine Learning.* NY: Academic Press, 1968.

Groves, T. and J. Ledyard. "Optimal Allocation of Public Goods: A Solution to the 'Free-Rider' Problem." *Econometrica*, v. 45(4), 1977: 783-809.

Groves T. and M. Loeb. "Incentives in a Divisionalized Firm." *Management Science*, v. 25(3), 1979: 221-230.

Holland, J.H. *Adaptation in Natural and Artificial Systems.* Ann Arbor, MI: Univ. Michigan Press, 1975.

Holland, J.H. Escaping brittleness: the possibilities of general-purpose learning algorithms applied to parallel rule-based systems. In R.S. Michalski et al., eds., *Machine Learning: An Artificial Intelligence Approach, v. II,* Los Altos, CA: M. Kaufmann, 1986: 593-624.

Holland, J.H. and J.S. Reitman. Cognitive systems based on adaptive algorithms. In D.A. Waterman and F. Hayes-Roth, eds., *Pattern-Directed Inference Systems,* NY: Academic, 1978: pp. 313-329.

Ichiishi, Tatsuro. *Game Theory for Economic Analysis.* NY: Academic, 1983.

Karlin, S. *Mathematical Methods and Theory in Games, Programming, and Economics,* Vols. 1 and 2. Reading, MA: Addison - Wesley, 1959.

Kim, S.H. Mathematical foundations of manufacturing science: theory and implications. Ph.D. thesis, M.I.T., May 1985.

Kim, S.H. A mathematical framework for intelligent manufacturing systems. *Proc. of Symposium on Integrated and Intelligent Manufacturing Systems,* ASME, Anaheim, CA, Dec., 1986, pp. 1-8.

Kim, S.H., "Difficult Problems and Creative Solutions", *International J. of Computer Applications in Technology*, v. 2(3), 1989: 171-185.

Kim, S.H. *Designing Intelligence: A Framework for Smart Systems.* NY: Oxford University Press, 1990.

Kim, S.H. "Coordination of Multiagent Systems through Explicit Valuation of Action." *Robotics and Computer-Integrated Manufacturing*, v. 8(4), 1991, pp. 265-291.

Kim, S.H. and N.P. Suh. Mathematical foundations for manufacturing. *J. Engineering for Industry*, v. 109(3), 1987: 213 - 218.

Kohonen, T. *Self-Organization and Associative Memory.* NY: Springer-Verlag, 1984.

Kompass, E.J. and T.J. Williams, eds. *Learning Systems and Pattern Recognition.* Barrington, IL: Technical Pub., 1983.

Lasdon, L.S. *Optimization Theory for Large Systems*. NY: Macmillan, 1970.

Marschak, J. "Elements for a Theory of Teams." *Management Science*, v. 1(2), 1955: 127-137.

Marschak, J. and R. Radner. *Economic Theory of Teams*. New Haven, CN: Yale Univ. Press, 1972.

McKinsey, J. *Introduction to the Theory of Games*. NY: McGraw-Hill, 1952.

Mendel, J.M. and K.S. Fu, eds. *Adaptive, Learning and Pattern Recognition Systems*. NY: Academic Press, 1970.

Minsky, M.L. *Computation: Finite and Infinite Machines*. Englewood Cliffs, NJ: Prentice-Hall, 1967.

Narendra, K.S. and M.A.L. Thathachar. Learning automata--a survey, *IEEE Trans. SMC*, v. SMC- (4), July 1974: 323-334.

Nof, S.Y. "Game Theoristic Models for Planning Cooperative Robotic Work." *Proc. NSF Design and Manufacturing Systems Conf.*, Austin, TX, Jan. 1991.

Owen, Guillermo. *Game Theory,* 2nd ed. NY: Academic, 1982.

Pearl, Judea. *Probabilistic Reasoning in Intelligent Systems*. San Mateo, CA: M. Kaufmann, 1988.

Radner, R. "Team Decision Problems." *Annals of Mathematical Statistics*, v. 33(3), 1962: 857-881.

Rajan, V.N. and S.Y. Nof. "A Game-Theoristic Approach for Co-operation Control in Multi-Machine Workstations." *Int. J. Computer Integrated Manufacturing*, v. 3(1), 1990: 47-59.

Rasmusen, E. *Games and Information*. NY: Basil Blackwell, 1989.

Sen, A.K. "Social Choice Theory." In K.J. Arrow and M.D. Intrilligator, eds., *Handbook of Mathematical Economics*, NY: Elsevier, 1986: 1073-1181.

Sen, A. and P.K. Pattanaik. "Necessary and Sufficient Conditions for Rational Choice under Majority Decision." *J. of Economic Theory*, v. 1, 1969: 178-202.

Shannon C.E. "The Mathematical Theory of Communication," *Bell System Tech. J.*, vol.27, 1948, pp. 379-423, 623-656.

Shannon C.E. and W.Weaver. *The Mathematical Theory of Communication*. Urbana, IL: U.of Illinois, 1949.

Shapiro, I. J. and K. S. Narendra. Use of stochastic automata for parameter self-optimization with multimodal performance criteria. *IEEE Trans. Systems Science and Cybernetics,* v. SSC-5 (4), Oct. 1969: 352-360.

Shubik, M. *Game Theory in the Social Sciences*. Cambridge, MA: MIT Press, 1982.

Tan, K.C. and R.J. Bennett. *Optimal Control of Spatial Systems*. London: George Allen & Unwin, 1984.

Tsypkin, Y.Z. *Adaptation and Learning in Automatic Systems,* trans. by Z.J. Nikolic. NY: Academic, 1971.

Valiant, L.G. "A Theory of the Learnable." *Communications of ACM*, v. 27(11), 1984: 1134-1142.

von Neumann, J. *Theory of Self-Reproducing Automata,* ed. by A.W. Burks. Urbana, IL: U. Illinois, 1966.

von Neumann, J. and O. Mergenstern. *Theory of Games and Economic Behavior*, 2nd ed. Princeton, NJ: Princeton Univ. Press, 1947.

Vorob'ev, N.N. *Game Theory*. Trans. by S. Kotz, NY: Springer-Verlag, 1977.

Young, T.Y. and K.-S. Fu, eds. *Handbook of Pattern Recognition and Image Processing*. NY: Academic, 1986.

Case Study in Coordination 4

The task of coordinating intelligent agents has become increasingly pervasive and significant. When autonomous agents such as robots are used in teams rather than as individuals, their coordination can become more critical for system performance than their individual capabilities. The deployment strategies and communication modes play an important role in the coordination of these teams.

This chapter examines a game theoretic deployment approach to robotic teams in an unstructured environment. A simulation model is developed and used to compare the performance of gaming rules with a non-anticipatory deterministic deployment rule. The initial game theoretic rule can be enhanced to exhibit both locally and globally adaptive characteristics. The new rule outperforms both the deterministic algorithm and the straightforward game-theoretic rule. This is achieved by adapting to trends in local regions in the environment as well as anticipating global eventualities.

Introduction

The coordination of a team of robots operating in an uncertain environment is related to various applications at the current stage of automation technology. This issue will assume even greater practical importance as automated equipment becomes more sophisticated. However, even at this juncture, we encounter applications of robots deployed in teams rather than individually.[1]

In the context of coordination, an important distinction between teams of people and of robots lies in the greater degree of uncertainty associated with human decision making activities. The overall efficiency of a human team depends in part on the efficacy with which individual preferences over outcomes of actions can be reconciled with team preferences. In a robot team this problem still exists since team members may be responsible for tasks external to the team as well as group-related objectives. However, individual preferences in a robotic team are generally immune to extraneous biases which often color human judgment.

---

[1] This chapter is based on Egilmez and Kim (1992).

Game theory has recently been identified as a promising tool with which robot teams can be analyzed and coordinated (Kim, 1991; Nof, 1991; Nof and Rajan, 1990). We must distinguish between human and robotic teams in the process of applying game theoretic concepts. For instance, equilibria of a game which are inefficient for human players may become efficient when robotic players are considered. Further, payoffs unattainable by the former may be available to the latter.

Fundamental differences between humans and robots in processing and sharing information can be used to exploit the assumption of rationality which underlies game theoretic approaches. The resulting solutions may be computationally effective but still not compelling enough for human players. Thus such concepts as Correlated Equilibrium (Aumann 1974, 1987) are better applicable to robotic players.[2] Further, in repeated games, the often simplifying assumption of perfectly patient players (corresponding to discount factors of unity) can be used with more confidence when the players are robotic. Other assumptions such as trust and observability of action are also more tenable for robotic players.

On the other hand, game theory is not a panacea for all problems involving robotic team coordination. Many interesting results from game theory arise from its application to competitive situations. One could even argue that team members must compete for resources while cooperating on a given task. Game theory may not always provide the proper models for such competition. When the resources are plentiful in relation to overall requirements, the competition will affect team efficiency only marginally.

This chapter approaches the robotic team coordination problem via a simulation model and uses analytical models to determine certain features of the simulation. The simulation model elaborates on – and helps to validate – the analytic model. The motivations for analytical models are as follows:

- Evaluation of various coordinative procedures.
- Comparison with results from the associated simulation.
- Development of design guidelines for robotic systems.

Analytical models

This section presents some analytical models that later serve as the basis for simulation work. The models incorporate representation modes as well as metrics for measuring temporal and spatial phenomena in the environment.

---

[2] For instance, the assumption of common prior probabilities is more justifiable when the players are robots.

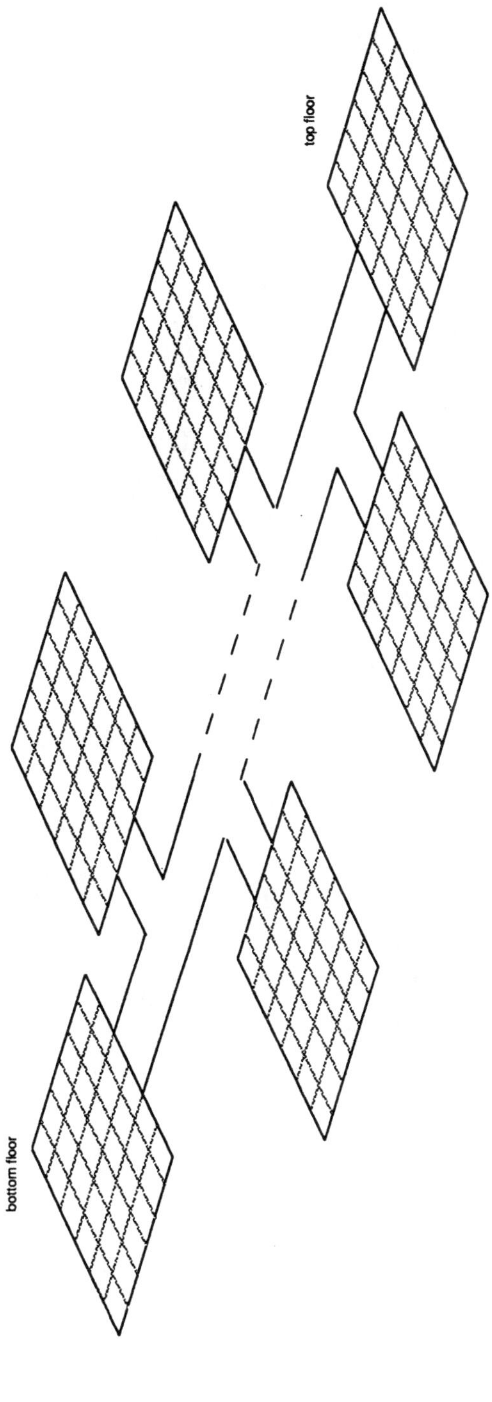

Figure 4.1b Generalization of the concept in Figure 4.1a to buildings with any number of floors. The elevator links in such buildings are modeled via corridors linking rooms in the two dimensional floor plan.

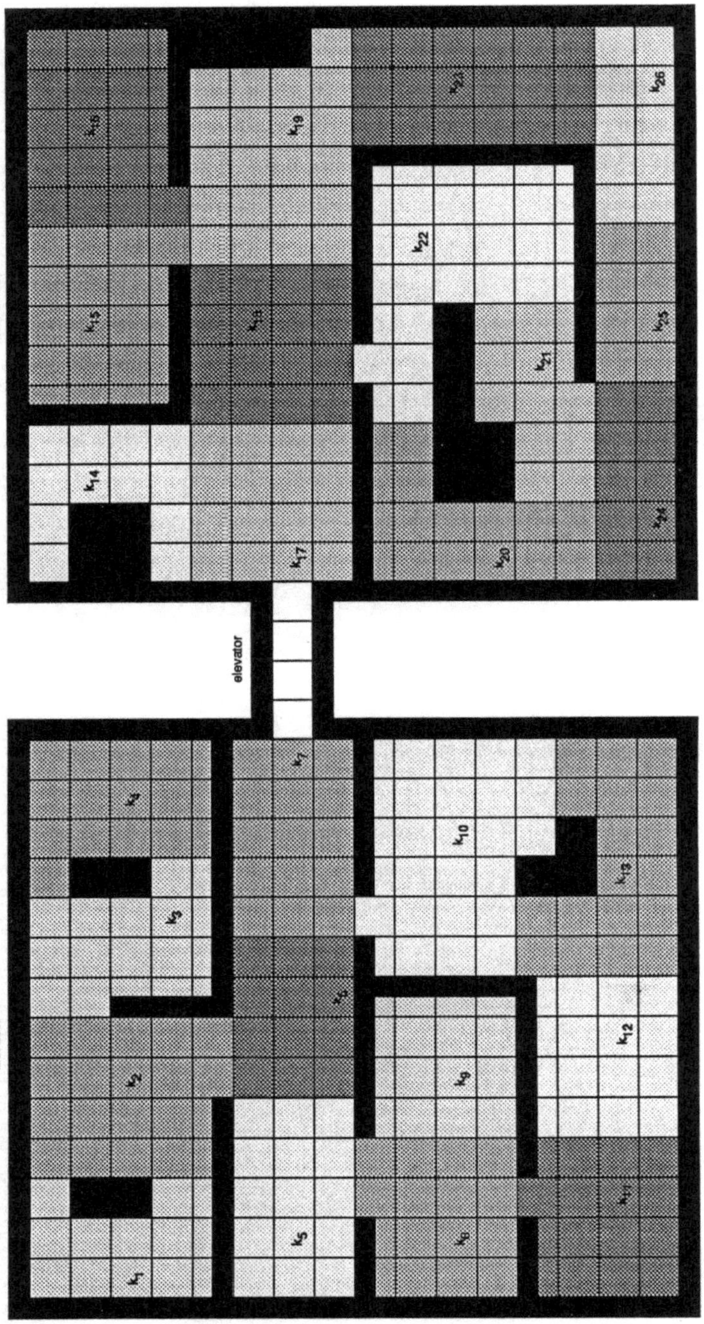

Figure 4.2 A sample environment. Regions are identified through shading. The capital cell associated with each region is also indicated.

Objects such as corridors – to represent elevators – are also regions. No two regions can overlap. That is, no two regions from $\boldsymbol{R}$ can contain the same cell from $\boldsymbol{C}$. The distance between two cells is given by the Euclidean distance in terms of their coordinates. That is, the distance between $p = (x_1, y_1)$ and $q = (x_2, y_2)$ is

$$\sqrt{(x_1 - x_2)^2 + (y_1 - y_2)^2}$$

for any two cells $p, q \in \boldsymbol{C}$.

An accessibility relation A_C is defined on $\boldsymbol{C}$. We call the set $\boldsymbol{M} = (\boldsymbol{C}, A_C)$ the *map* of the environment. This map defines the free cells in the environment and their connectivity. More formally, we can write the accessibility relation for all cells in $\boldsymbol{C}$ as

$$(x_1, y_1) \mathrm{A_C} (x_2, y_2) \Leftrightarrow \sqrt{(x_1 - x_2)^2 + (y_1 - y_2)^2} < 2$$

$$\text{such that } \exists\, r_1, r_2 \in \boldsymbol{R} \text{ with } (x_1, y_1) \in r_1 \text{ and } (x_2, y_2) \in r_2.$$

$\mathrm{A_C}$ is both reflexive and symmetric. In other words, two cells in $\boldsymbol{C}$ are normally accessible from each other if and only if they are either edgewise or diagonally adjacent and both are free.

The relation A_C defines the correspondence $A_{\boldsymbol{R}}$, which is an accessibility relation defined on $\boldsymbol{R}$. That is, for any two regions r_1 and r_2 in $\boldsymbol{R}$

$$r_1\, A_{\boldsymbol{R}}\, r_2 \Leftrightarrow c_1\, A_C\, c_2 \text{ for some } c_1, c_2 \text{ such that } c_1 \in r_1 \text{ and } c_2 \in r_2.$$

In other words, two regions are accessible precisely when there are cells in both that are accessible from each other.

A *capital* cell is associated with every region of the environment. This cell represents a possible target location for a robot as part of its assigned duties. Let $\boldsymbol{K}$ denote the set of all capital cells for the environment. When a robot is asked to respond to an event occurring in some region of the environment, it takes the capital cell of that region as its goal.

Given the map $\boldsymbol{M}$ and the distances between cells, we can calculate the shortest route from every free cell in $\boldsymbol{C}$ to each capital cell in $\boldsymbol{K}$. A dynamic program can be used for this purpose; an example is a *label correction* algorithm (Denardo and Fox 1979; Tsai et al., 1990). Considerable computational efficiency can be achieved by taking into account the structure of the environment (Dreyfus 1969). Such an algorithm can be used to define a distance function d:

$$d : \{\text{free cells in } \boldsymbol{C}\} \times \boldsymbol{K} \rightarrow \boldsymbol{R}$$

where $d(c, k)$ is a function that gives the length of the shortest path between cell $c \in r$, for some $r \in \mathbf{R}$ and capital cell $k \in \mathbf{K}$. This is the length of the shortest path between the two cells according to what is known in $\mathbf{M}$ about obstructions in the environment. Thus a change in the map will change this value.

The algorithm also defines the path p:

$$p: \{\text{free cells in } \boldsymbol{C}\} \times \boldsymbol{K} \rightarrow 2^{\mathbf{C}}$$

Here, 2^C defines the power set of $\boldsymbol{C}$. Consequently, the relation $p(c, k)$ identifies the nominal shortest path between free cell c and capital cell k. If $p(c, k)$ contains capital cells other than k we say that the path has subgoals. A path with subgoals can be split into its components. For example,

$$p(c, k_2) = p(c, k_1) \cup p(k_1, k_2) \;\; \text{if} \;\; k_1 \in p(c, k_2)$$

This follows from the principle of optimality.

A path is called *restricted* if it passes through a *restricted region.* A restricted region is one which limits the number of its normally free cells that can be occupied simultaneously. Let $\boldsymbol{R}_r$ denote the set of restricted regions. Then $p(c, k)$ is restricted if for some $c_1 \in p(c, k)$, there is a region $r \in \boldsymbol{R}_r$ for which $c_1 \in r$. In contrast, when such a path exists, $p^*(c, k)$ is called an *unrestricted alternate path* from c to k. Restricted regions can be used to model tight corridors, passageways, doorways, etc.

A special kind of region in the environment is one that models an elevator. The set of all elevator regions is denoted by $\boldsymbol{R}_e$. If c and k are on two different floors of a building, then $p(c, k)$ must necessarily pass through an elevator region. An alternate path will exist only if the two floors are connected by more than one elevator. In our environments, corridors of single cell width will be used to model elevators. Thus defining elevators as restricted regions is not feasible when an elevator can carry more than one passenger. The set of regions other than elevator regions is denoted by $\boldsymbol{F} = \boldsymbol{R} - \boldsymbol{R}_e$.

Time to Respond to an Event

The average response time (ART) for a robot responding to an event at capital cell k when it is located at cell c depends on the speed with which the robot can travel in the environment as well as the traffic conditions prevailing in the environment during the trip. Traffic congestion can spring from delays caused by humans and other robots moving in the environment and thereby blocking shortest routes. In addition, delays can be caused by having to wait for elevators. Thus

$$\text{ART}(c, k) = \frac{d(c, k)}{v_R} + \Delta_E(c, k) + \Delta_T(c, k)$$

where

v_R = speed of the robot
Δ_E = average delay due to elevator if c and k are on different floors
Δ_T = average delay due to traffic in the environment

ART is given by the sum of three components as shown above. The first component gives the nominal delay if the shortest path between c and k were taken and no additional delays were encountered. For simplicity's sake, we will henceforth assume a fixed speed of unity for v_R. In general, cell sizes can be adjusted to model different speeds in distinct regions of the environment. With unity value for v_R, we have

$$\text{ART}(c, k) = d(c, k) + \Delta_E(c, k) + \Delta_T(c, k)$$

The delay due to elevator usage and traffic on the route between c and k are random variables. Consequently we will focus on their average contributions to the overall response time.

Elevator Delays (Δ_E). When an elevator must be used, the average delay en route from c to k can be decomposed into the average waiting time for an elevator plus the average travel time. The average travel time is modeled in our environments by the length of the single-cell corridor representing the elevator link between floors. This aspect is included in $d(c, k)$. The average waiting time for elevators depends on such factors as the passenger arrival rate, elevator capacity, speed, and so on. It is possible to model a general multi-car elevator system as a multi-channel bulk service queueing system (Alexandris et al., 1979a, b; 1981). The average waiting time can then be found analytically given the following assumptions about the system.

- A majority of the passengers board elevators on the ground floor and go up.
- Elevators do not leave without passengers.
- Elevators return to the ground floor when they become free.
- The passenger waiting queue capacity is not limited.
- Passengers arrive at the ground floor at a fixed Poisson arrival rate λ.
- Passengers have no preferences among different elevators.
- Waiting passengers board on a first-come/first-served basis.
- Elevator carrying capacity is not exceeded at any time.
- The batch service time distribution is exponential with mean μ^{-1} = RTT. RTT is the Round Trip Time; that is, the time for a single trip up and down the building on one elevator.

These assumptions lead to a worst case estimate for the average waiting time. Empirical studies of multi-car elevator systems yield performance characteristics that have been shown to correlate well with those found from the preceding model (Alexandris et al., 1979a, b; 1981).

The average passenger waiting time (AWT) according to the above queueing model is

$$\text{AWT} = \frac{q}{\lambda}$$

where q is the average queue size and is given by (Alexandris et al., 1979a)

$$q = \frac{\alpha^L (Z_0 - 1)^{-1}}{(L\alpha)^L + L!\,(1 - Z_0^{-1}) \sum_{r=0}^{L-1} \frac{\alpha^r}{r!}}$$

The parameter L denotes the number of elevators in the system, $\alpha = \lambda/\mu$, and Z_0 is the root outside the unit circle from the equation

$$\rho m Z^{m+1} - (\rho m + 1) Z^m + 1 = 0$$

Here m is the elevator carrying capacity, ρ is the utilization factor of the system, and ρ must be less than 1 for stable operation. The parameter ρ is given by the ratio of passengers carried in an elevator to its carrying capacity:

$$\rho = \frac{\lambda}{\mu m L} = \frac{\alpha}{mL}$$

The theoretical model above as well as empirical results indicate that at utilization levels exceeding $\rho = 80\%$, system performance declines rapidly (Alexandris et al., 1979a, b; 1981).

These results can be integrated into our model by computing AWT for a given environment and then setting $\Delta_E(c, k) = \text{AWT}$ for all c and k for which an elevator must be used. The queueing model results suggest that when computing AWT for an environment (e.g. for robotic deployment in an office building) we can approximate various elevator utilization levels by varying ρ. This can be done for a given environment with fixed m and L by varying α, the traffic intensity of that environment.

The example environment in Figure 4.2 has a single car ($L = 1$), single capacity ($m = 1$) elevator. Thus

$$q = \frac{\alpha (Z_o - 1)^{-1}}{\alpha + (1 - Z_o^{-1})}$$

and $Z_o = \rho^{-1}$. Therefore

$$\text{AWT} = \frac{1}{\mu}\left(\frac{\rho}{1 - \rho}\right)$$

Figure 4.3 plots AWT for this elevator system as a function of ρ and μ. From the figure we see that at utilization levels close to 100% the average waiting time for the elevator is unbounded. The round trip time μ^{-1} for the elevator in Figure 4.2 is 8 units. Taking $\rho = 0.8$ we obtain AWT = 32 as a realistic worst case estimate for this environment.

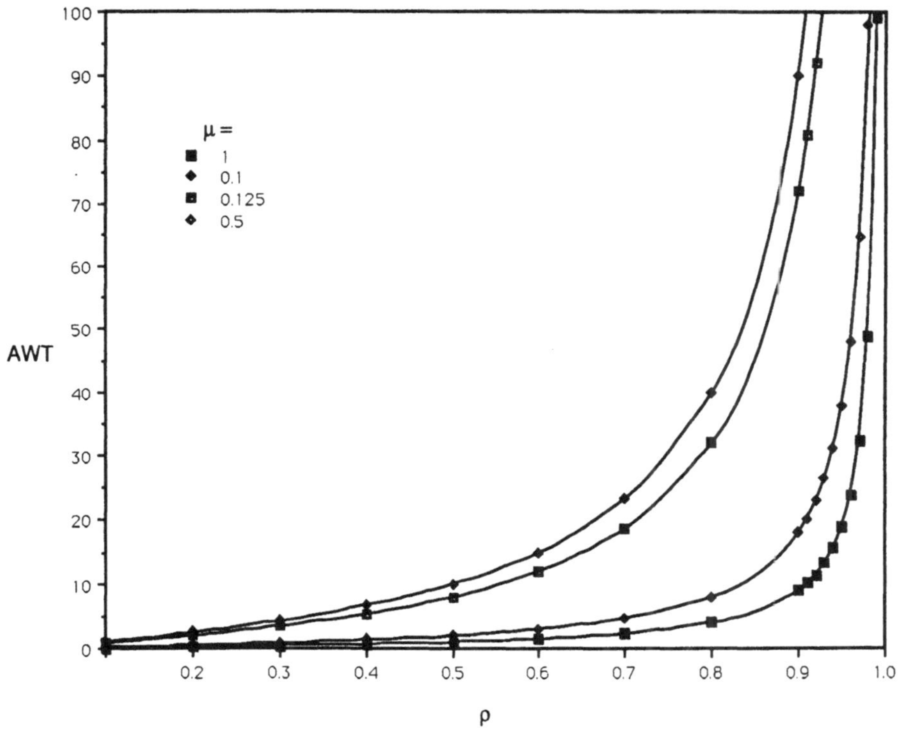

Figure 4.3 Average waiting time (AWT = q/γ) for a single car, single capacity elevator as a function of its utilization (ρ) and service rate (μ).

Traffic Delays (Δ_T). A robot can be delayed en route due to the movement of other entities, human as well as robotic. Clutter in the environment which blocks normally open routes can also cause delays. The average traffic delay for a portion of the environment (e.g. a floor in an office building) can be analyzed using the notion of network flow. The nodes of such a network are the regions in that portion of the environment and the links represent $\boldsymbol{A_R}$, the accessibility among regions. In network flow analysis, each link in the network is assigned a capacity. In our case this link capacity will be given by the maximum number of simultaneous crossings possible between the two regions connected by the link. Formally, the capacity for the link between regions r_1 and r_2 is given by

$$c_{12} = c_{21} = |\{(c_1, c_2) : c_1 \in r_1, c_2 \in r_2, c_1 \boldsymbol{A_C}\ c_2\}|$$

where $|\{\bullet\}|$ denotes the cardinality of the set $\{\bullet\}$.

The maximum flow between each pair of regions can then be calculated. The greatest flow between two regions will define the maximum number of travelers supported in parallel by the connecting route.

Figure 4.4 illustrates the application of network flow analysis to the example environment in Figure 4.2. Let the following mapping represent the maximal flows calculated for a set of regions.

$$f : F \times F \rightarrow \mathbb{N}$$

For instance, $f(r_1, r_2)$ returns the maximum possible congestion-free traffic between regions r_1 and r_2.

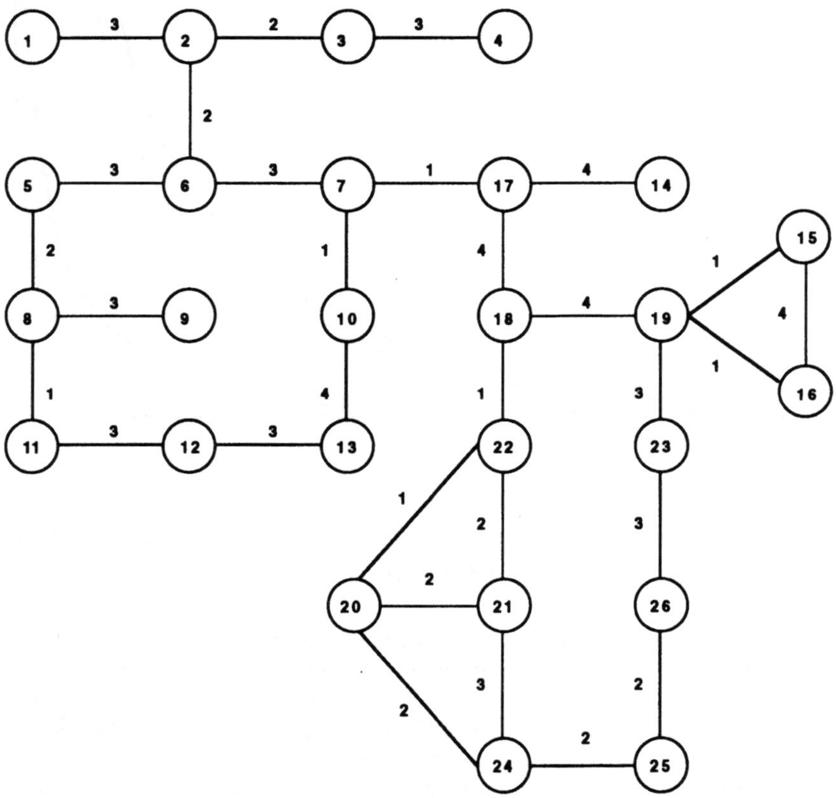

Figure 4.4 Flow capacity graph of the regions of the environment of Figure 4.2. This graph is used to estimate traffic congestion in the environment.

The delay due to one unit of excess traffic flow between two regions is assumed to be one time unit. To illustrate the rationale for this assumption, imagine three entities traveling along a route between two regions. Suppose the maximum traffic flow between these two regions allows only two entities to travel abreast. Then the third entity can only traverse the same route by staying behind the previous two, and the best it can do is to stay one time unit behind them. If there were four robots rather than three, the second two could follow immediately behind the first pair, thereby resulting in a delay of 1 time unit as before.

More generally, let N denote the number of agents (humans and robots) operating in the environment. Then the worst case traffic delay between two regions is given by

$$\Delta_T(r_1, r_2) = \text{ceil}[\, N \,/\, f(r_1, r_2)] - 1 \qquad : N \geq 1, f \geq 1.$$

Here ceil[x] denotes the ceiling of x; namely, the smallest integer larger than or equal to x. The traffic delay between individual cells is then given by

$$\Delta_T(c, k) = \Delta_T(r_1, r_2) \text{ such that } c \in r_1 \text{ and } k \in r_2.$$

Simulation model

A simulation program called Robot Deployment Simulator (RDS) has been developed in Smalltalk/V for the Macintosh computer. The program consists of two parts: an environment definition module and a simulation/animation module.

Figure 4.5 shows the environment definition module in action. Here the environment in Figure 4.2 is shown as it was entered using the Environment Editor window of RDS. This window provides a graphical editor for drawing environment floor plans. A grid of a specified dimensionality is superimposed on the window for the user's convenience. The convention for specifying multi-level environments is as described previously in connection with Figure 4.1.

Once an environment is entered, its map $\boldsymbol{M}$ is compiled and stored internally. The user then specifies the capital cells. Subsequently all paths originating from free cells and having these capital cells as their destinations are determined. These paths and their associated lengths are stored as a tree structure. The environment editor also constructs a three dimensional rendering of the environment floor plan and stores it for future use by the animation module.

The program maintains multiple environments in an internal database. The primary purpose of the environment editor module is to quickly specify or call up environments or the floor plans of segments of an environment, and to edit them if necessary. Thus one can investigate, for instance, the effects of static obstacles on the performance of a robot team. Algorithms for both static and dynamic obstacle avoidance can thus be tested through several modifications of the same general environment.

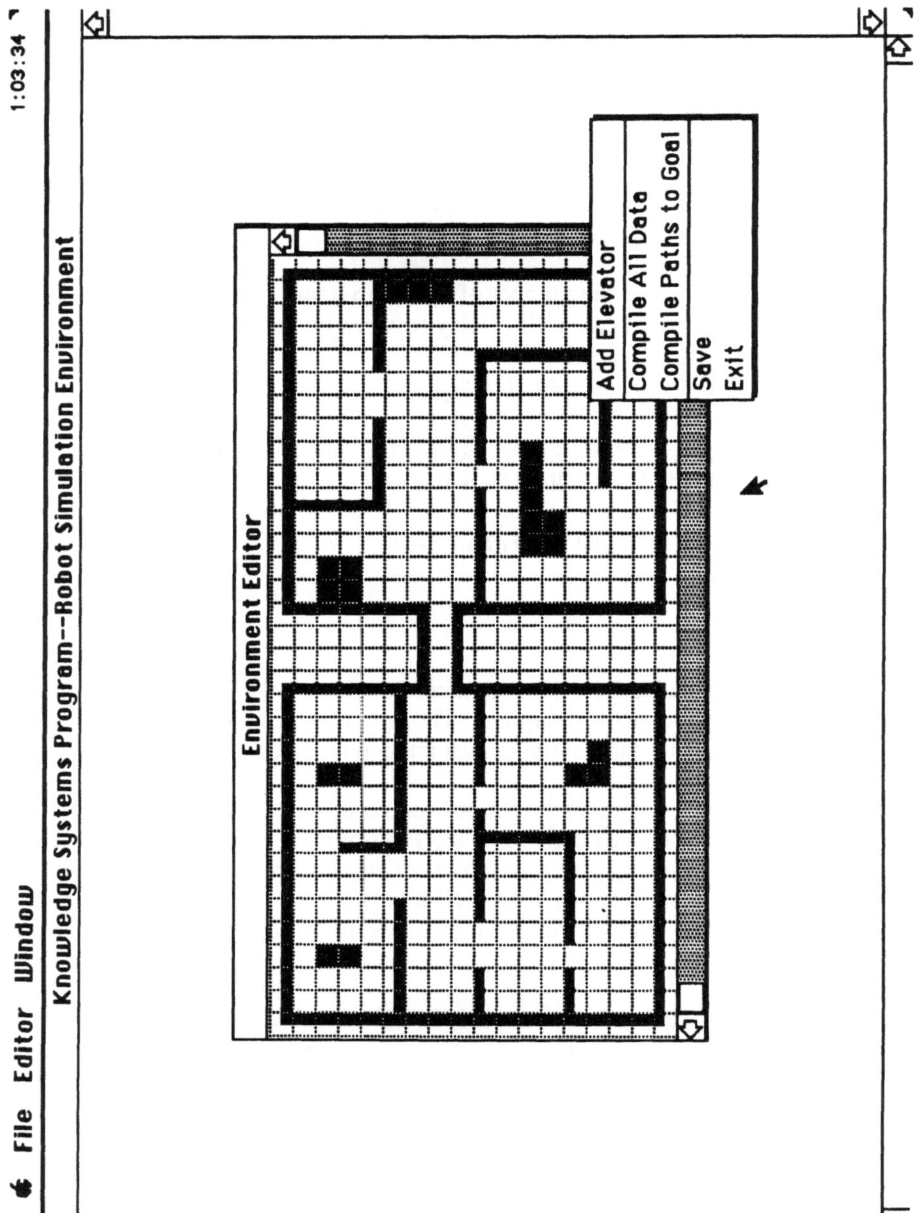

Figure 4.5 Environment Definition Window of RDS

Figure 4.6 shows the simulation window of RDS. Here we have a partial three dimensional view of the environment of Figure 4.2. This view was generated when the environment defined in the editor module was compiled. This window is the primary interface between the user and the program. Here one can generate random sequences of events with specific (Poisson) arrival rates for subsequent use as input to a robot deployment simulation. Any number of human and robot agents can be specified and one of several predefined deployment rules can be used to simulate the response of the resulting robotic team.

Human movements are generated by randomly selecting capital cells as destinations. The program can be instructed to record these movements so that subsequent simulations are run under the same traffic conditions. The simulations incorporate traffic congestion due to human movements and a dynamic obstacle avoidance algorithm for the robots. Consequently, various obstacle avoidance algorithms can be embedded in the simulation to test their relative performance.

Another source of travel delay in the environment springs from the elevators. A separate object is used to simulate the use of elevators. This object keeps a first-in/first-out (FIFO) request queue and uses it to order human and robotic travel among floors. In brief, the simulation supports three dynamic objects and one static object – namely robots, humans, elevators, and the environment.[3]

Using objects to model agents and various aspects of the environment adds realism to the simulation: each object–be it human or robot–can be programmed with behavior at arbitrary levels of detail. For example, although the elevator object processes requests rather simply in the current application, in other situations the user could program various request processing rules to evaluate their efficacy.

RDS was not intended as a tool to measure the run-time performance of various algorithms. For this reason, all relevant calculations are completed at each step before the simulation continues. The tacit assumption is that all algorithms execute in real-time in comparison to the simulation step size.[4]

Figures 4.7 and 4.8 show snapshots of a typical simulation run. The former figure depicts two robots and two humans at their initial positions. The elevator is parked at the first floor in readiness for service. Figure 4.8 shows a time lapse screen dump of the first five steps of the simulation run. An event occurs on the first floor and robot #2 is dispatched to it. Two time steps later another event, this time on the second floor, occurs and robot #1, the only other free robot, is dispatched. The humans are shown moving in the environment according to their own agendas.

---

[3] A possible extension for the future is to include other stationary or mobile objects (e.g. beds or wheelchairs in a hospital environment).

[4] The program can be modified to employ semaphores for modeling the run time of algorithms during obstacle avoidance or robot deployment. This is a relatively simple task in Smalltalk/V, where semaphores are standard objects.

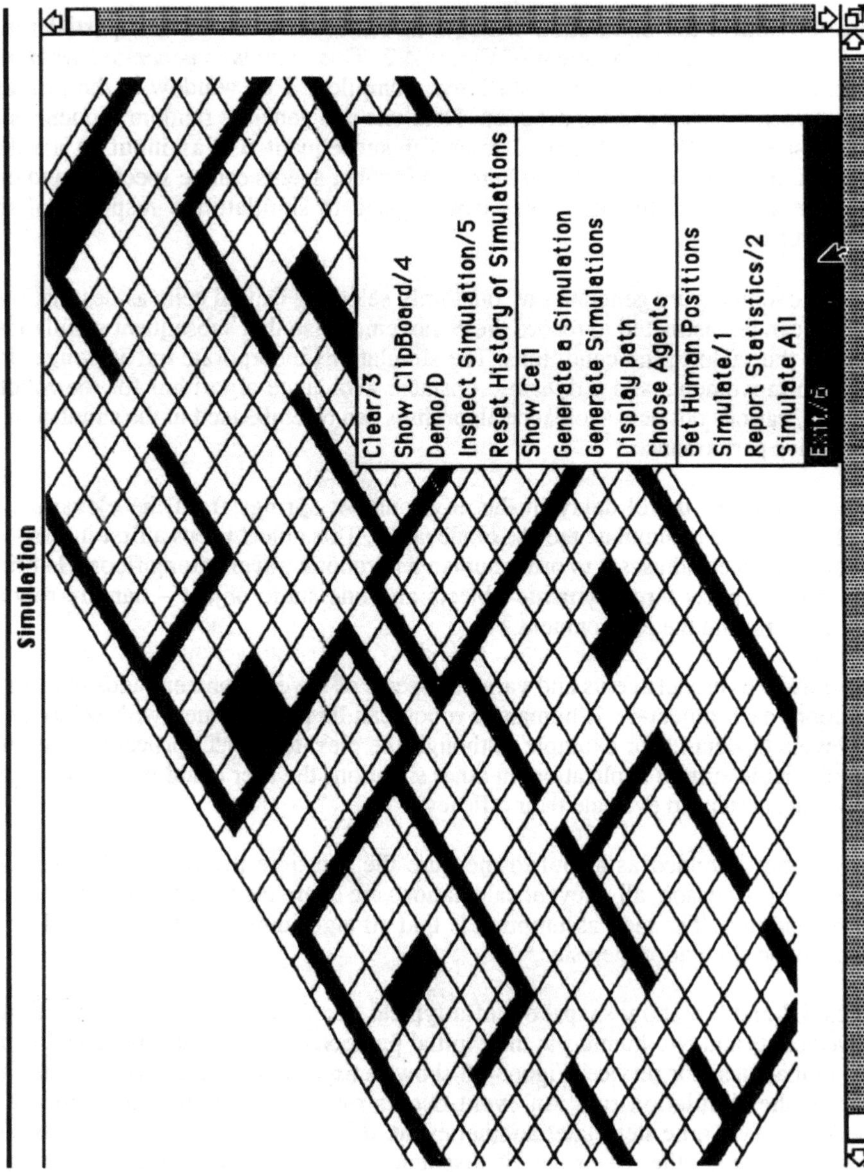

Figure 4.6 Simulation window of RDS showing a three dimensional view of the environment from Figure 4.2. A pop-up menu is shown indicating the commands available to the user.

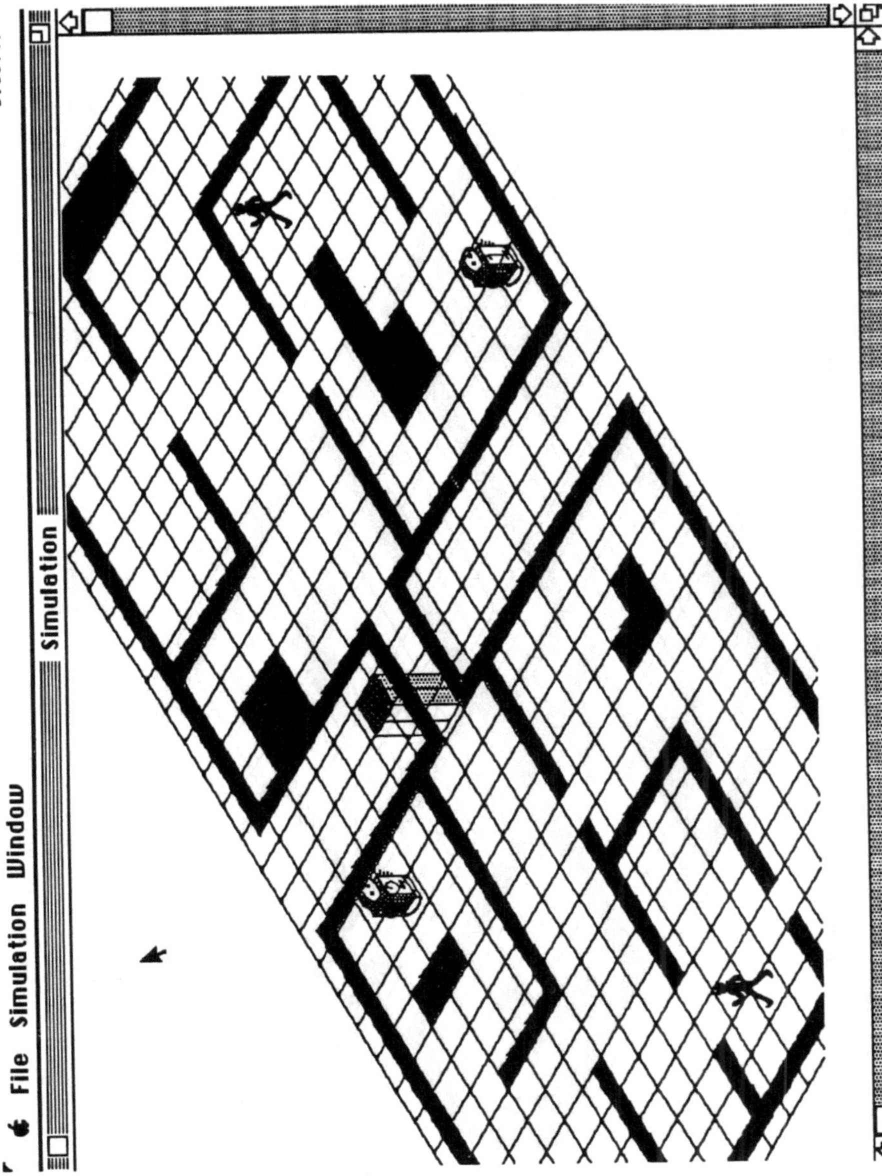

Figure 4.7 Simulation window of RDS. Shown ar e agent starting positions for a sample simulation.

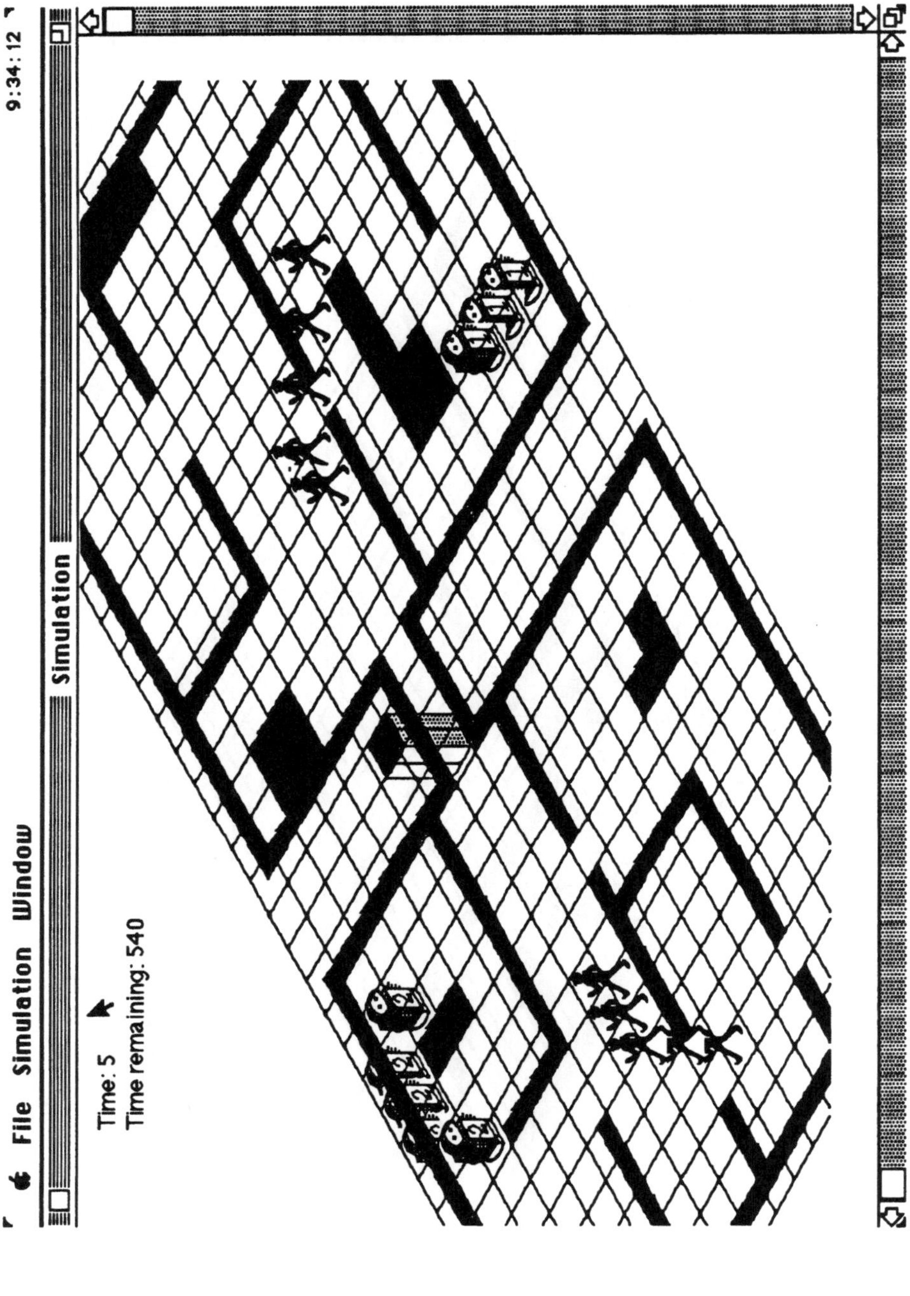

Figure 4.8 Status of sample simulation at 5 time units.

The next section will discuss the rules for positioning free robots in the environment and dispatching them in response to events. The simulation keeps a FIFO wait queue for events which occur when all robots are busy. The response time for each event, and the amount of time each event spends in the wait queue, are recorded.[5]

Simulation Specifics

The environment of Figure 4.2 was used to run a number of simulations with varying robot team sizes and deployment rules. Events were generated assuming Poisson arrival rates. First, 100 event positions were generated with locations sampled from a uniform distribution over the 26 capital cells shown in Figure 4.2. Several simulation runs were then generated using this sequence by varying the arrival rate, equal to the reciprocal of the mean of the exponential interarrival time distribution. Thus by increasing the arrival rate it was possible to simulate increased system loading.

Traffic conditions due to human movements should be kept uniform among simulation runs. Consequently, a sequence of goal locations were generated for each human object. Human agents were instructed to move to the next location in such a sequence once they had arrived at their current goal location. The same sequences were used for each simulation run. The simulation setup described so far was designed to test the performance of different deployment rules as more events occurred at shorter time intervals. The locations of events were kept constant across trials to ensure that each deployment rule faced the same occurrence pattern for a given arrival rate.

The performance of a given deployment rule under various loading conditions was measured using the mean response time and the variance of the response times to the 100 events in each trial, as well as the arrival rate of events (λ). A low mean response time indicates good performance. However, low variation of response times over the 100 events is also important. Thus an appropriate performance measure is:

$$\eta = \log\left(\frac{\lambda^{-3}}{\mu \times \sigma^2}\right)$$

[5] Although it was not implemented, the elevator object can be programmed to record delays accrued in responding to requests for transport. This data can be used to calculate its utilization level as well as the average delay that a human or robot can expect when it wishes to use the elevator. Thus Δ_E can be learned rather than estimated using the analytical model described previously. Rules can then be designed whereby the elevator agent decides how to respond to requests for transport based on this learned Δ_E rather than the FIFO approach currently employed. When more elevators and floors are involved, coordination among different elevator agents may also be an important topic for investigation. The goal of the present investigation, however, was to compare robotic team deployment rules. The simulation model was designed with maximum flexibility in mind to accommodate diverse types of inquiry in the future.

where

$$\mu = \frac{1}{n} \sum_{i=1}^{n} y_i$$

and

$$\sigma^2 = \frac{1}{n-1} \sum_{i=1}^{n} (y_i - \mu)^2$$

are the sample mean and sample variance, respectively. The y_i's are the observed response times to events. For the simulations discussed here, $n = 100$.

This performance measure is motivated by the fact that if the ratio of the mean time between events (λ^{-1}) to the observed mean response time (μ) falls below 1 we can conclude that the system is being overloaded. At this point, the rate at which events can be serviced is lower than the rate at which events occur. A similar meaning can be attached to the ratio of the variance of the interarrival times (λ^{-2}) to the variance of the response times (σ^2). Thus negative values for η can be interpreted as indicating system saturation or overloading.

We can also write the performance measure as

$$\eta = \log\left(\frac{\lambda^{-1}}{\mu}\right) + \log\left(\frac{\lambda^{-2}}{\sigma^2}\right)$$

Thus the overall performance metric is the simple sum of contributions due to the relative mean and the variance of response times.

Deployment Rules

A number of previous studies on the dispatching of Automated Guided Vehicles (AGVs) have tested a variety of dispatch rules, primarily through simulation studies.[6] Results from these studies indicate that the performance of different rules tend to converge as the system load is increased (in our case, as the interarrival rate is increased). To date, however, rules that anticipate demand have not been adequately tested. Results from previous studies suggest that, when demand is not anticipated, the performance of different dispatch rules does not vary significantly. Here we will see that when demand is anticipated in the form of a game theoretic rule, performance is significantly improved, even under severe system loading conditions.

---

[6] Egbelu (1986); Cheng (1987); Egbelu and Tanchoco (1987).

Nearest Robot Rule (NRR). For this procedure, an available robot closest to an event responds to it. When the robot arrives at the event location, it remains there until it is dispatched once again. In the simulation, an event is considered to be resolved once the robot dispatched to it arrives there.[7]

The NRR deployment rule was used as a benchmark for comparing game theoretic rules. Since the NRR does not anticipate demand in the form of next event locations, one would expect it to perform more poorly than the game theoretic rules which do look ahead.

Shapley Value Solution Rule (SVR). As explained in Chapter 3, the Shapley Value solution to an *N*-person game in characteristic function form reflects the bargaining power possessed by each player (Rapoport, 1970). That is, the Shapley value of the game for a player indicates the expected value which that player adds to a coalition. The characteristic function form of a game assigns a value to all possible coalitions of the players in the game.

A straightforward approach to modeling the deployment problem in game-theoretic terms is to view the robots as players. This is only natural since the robots apear to be the entities involved in a multiagent decision problem. Moreover, these entities must cooperate (i.e. form coalitions) so that they can respond to events efficiently as a team. They must also compete for resources such as space and elevators while responding to events.

When we approach the deployment problem with the robots as players in a collaborative game, however, we find that a utility structure must be established to incorporate the cooperative and competitive factors. The attempt metamorphoses into one of solving a dynamic constrained optimization problem with a team cost function. Here we take another approach in order to retain the game theoretic flavor of the situation.

Figure 4.9 shows a graph model of the environment from Figure 4.2. The nodes represent capital cells and their associated regions. The links represent routes between adjacent regions. The link weights give the average delay a robot experiences on that route. In general we may have *N* regions.

Resources and their sharing are modeled through the link weights. Thus we employ ART(k_1, k_2) to compute the weight on the link between regions 1 and 2. Each node is weighted in terms of the likelihood of an event occurring in that region. Thus these weights model demand anticipation. The context provides opportunity for adaptation and learning. We define a game of *N* players with the ***regions*** in the environment as the ***players***. The regions have *S* available robots that they can distribute among themselves. In general $N \geq S$ so that some subset **S** with cardinality *S* among the *N* regions must cooperate by distributing the available robots among themselves. The value to a coalition **S** is determined by considering the coverage obtained when the available robots are

---

[7] A simple extension is to instruct each robot to spend a random amount of time at the event location once it arrives. In addition, a robot might be sent to another — perhaps randomly chosen — location in the environment after arrival at the initial destination. These modifications to the current simulation would incorporate randomness in both the service time and waiting locale.

positioned in the member regions. Note that as events occur and robots are dispatched the number of available robots will change.

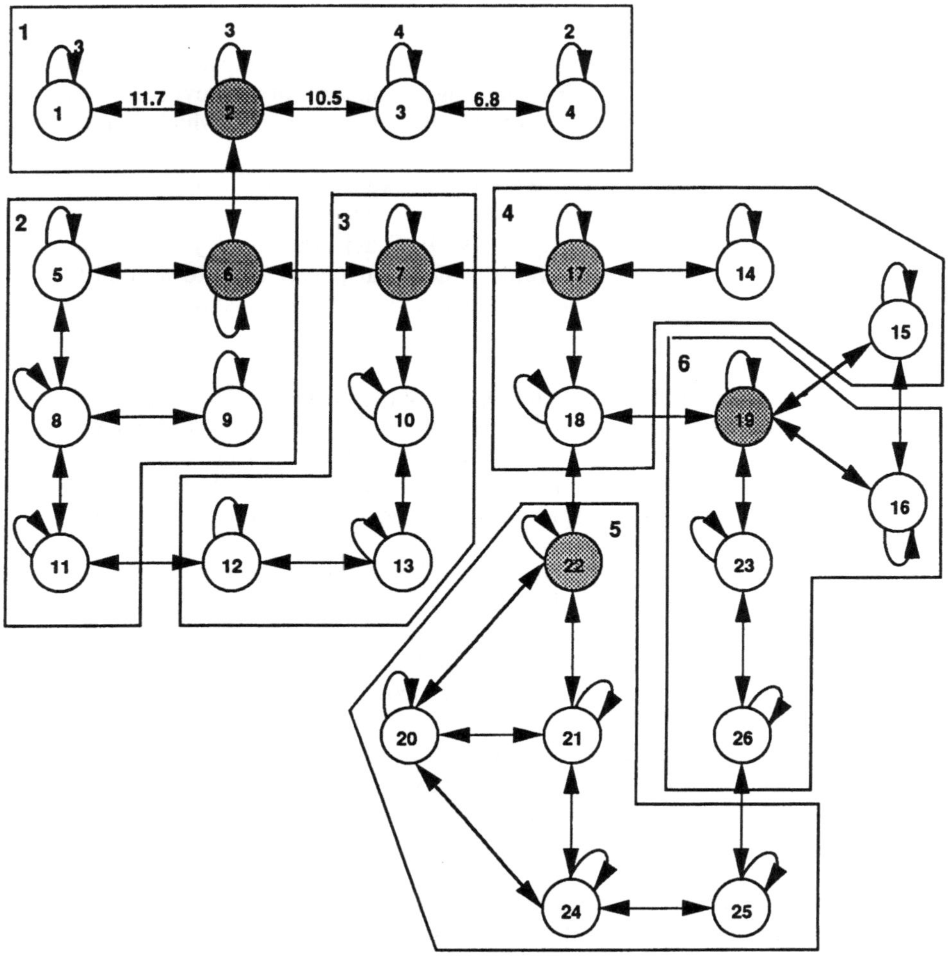

Figure 4.9 Graph model of the environment from Figure 4.2. The delays between regions of subgame 1 are shown. Delays for other regions are similar. The six subgames are identified and the winner region in each one is highlighted.

The following defines the value of the game for coalition **S**:

$v(\mathbf{S})$ = reciprocal of the average delay to an event for coalition **S**.

More formally

$$v(\emptyset) = 0.$$

$$v(\mathbf{S} \in 2^{\mathbf{N}}) = (\mathbf{w}^{\mathrm{T}}\, \mathbf{d}(\mathbf{S}))^{-1}$$

where **N** is the set of regions, $2^{\mathbf{N}}$ the power set of **N**, **w** the likelihood vector for event occurrences, and **d**(**S**) the vector of minimum delays on routes from regions in **S** to all regions in **N**. The notation $\mathbf{a}^{\mathrm{T}}$ denotes the transpose of vector **a**, where all vectors are taken to be column vectors. Here **w** is a probability distribution over the regions of an environment; the ith component of **w** gives the probability of an event occurring in region i. The components of **d**(**S**) are computed as follows

$$d_i\,(\mathbf{S}) = \min\,\{\ \mathrm{ART}(k_j, k_i) : \forall j \in \mathbf{S}\}$$

In other words, the i th component of **d**(**S**) is the minimum delay from a region in **S** to region i. Thus v(S) returns the coverage that S = |S| robots provide when they are placed at the capital cells of the regions in S.

The Shapley value to region $i \in \mathbf{N}$ is then given by

$$\varphi_i = \sum_{\mathbf{S} \in 2^{\mathbf{N}}} \frac{(S\ -\ 1)!\ (N\ -\ S)!}{N!}\,[v(\mathbf{S})\ -\ v(\mathbf{S}\ -\ \{i\})]$$

where, as indicated, the summation is over all subsets of the set of regions. The regions in the environment can then be ranked in terms of their Shapley values. A region with a high value of φ_i represents a better strategic location for positioning an available robot than others with lower values.

For the environment in Figure 4.2, we have 26 regions. This creates computational problems when each region's Shapley value must be calculated. In general the computational complexity of finding Shapley values for N regions grows as $O(2^N)$. For even moderate values of N, this is costly. Further, it is unlikely that N robots will be available for deployment in an N-region environment. Consequently, it seems worthwhile to consider subgames among subsets of **N**. We will later see that such an approach becomes even more worthwhile as larger robot team sizes begin to create traffic congestion problems. In that case, redeployment is necessary after a change in the number of available robots.

Figure 4.9 identifies 6 superregions, each of which consists in turn of 4 or 5 regions. Subgames can be defined for each of these superregions. The above formulas for characteristic values of coalitions and Shapley values remains unchanged if the collection **S** is now taken to be a subset of the set of regions constituting a superregion. Thus the value to a coalition still represents the coverage over the entire environment, but now only when robots are placed in regions of a superregion rather than the whole environment. The rank ordering among the regions of a superregion, based on the Shapley value, can be used to decide which capital cell should hold a robot when the latter becomes free in the superregion containing that capital cell.

SVR takes the above approach one step further and plays a supergame among the winners of the subgames. More specifically, each player of the supergame is the region which obtained the highest Shapley value in its subgame; that is, the best region for covering the whole environment when a robot was available for placement in that superregion.

When SVR is applied to the environment of Figure 4.2, a rank ordering is obtained among 6 of the 26 regions. This means that a team of at most 6 robots can be deployed in this environment using the following rule: when $n \leq 6$ robots are free, deploy them to the first n regions of the rank ordering obtained in the above manner. The results of the subgames and the supergame played over the 26 regions of this environment are as follows (the locations of capital cells are given in Figure 4.2).

The first interaction, subgame 1, involves superregion 1 which consists of regions 1, 2, 3, and 4. The Shapley vector for this superregion is given by

$$\varphi_1 = 0.0044775,\ \varphi_2 = \mathbf{0.0068671},\ \varphi_3 = 0.0045716,\ \varphi_4 = 0.0040069.$$

Subgame 2 among regions 5, 6, 8, 9, and 11 (superregion 2) gives

$$\varphi_5 = 0.0042103,\ \varphi_6 = \mathbf{0.0072488},\ \varphi_8 = 0.0038733,\ \varphi_9 = 0.0038064,$$

$$\varphi_{11} = 0.0037003.$$

Subgame 3 among regions 7, 10, 12, and 13 (superregion 3) generates

$$\varphi_7 = \mathbf{0.0093453},\ \varphi_{10} = 0.0057949,\ \varphi_{12} = 0.0051808,\ \varphi_{13} = 0.0048268.$$

Subgame 4 among regions 14, 15, 17, and 18 (superregion 4) leads to

$$\varphi_{14} = 0.0054942,\ \varphi_{15} = 0.0048562,\ \varphi_{17} = \mathbf{0.0084003},\ \varphi_{18} = 0.0065882.$$

Subgame 5 among regions 20, 21, 22, 24, and 25 (superregion 5) gives

$$\varphi_{20} = 0.0039478,\ \varphi_{21} = 0.0042277,\ \varphi_{22} = \mathbf{0.0053289},\ \varphi_{24} = 0.0034671,$$

$$\varphi_{25} = 0.0037142.$$

Subgame 6 among regions 16, 19, 23, and 26 (superregion 6) yields

$$\varphi_{16} = 0.0048764,\ \varphi_{19} = \mathbf{0.0059339},\ \varphi_{23} = 0.0047449,\ \varphi_{26} = 0.0043125.$$

The winner in each subgame is highlighted in bold.

The winning regions in turn constitute players in a supergame. The Shapley values for the supergame among regions 2, 6, 7, 17, 19, 22 are

$$\varphi_2' = 0.013925,\ \varphi_6' = 0.016052,\ \varphi_7' = 0.010177,\ \varphi_{17}' = 0.012202,$$

$$\varphi_{19}' = 0.016331,\ \varphi_{22}' = 0.015446.$$

The Shapley variables of the supergame are distinguished from their subgame counterparts by the use of primes.

SVR uses the results of the foregoing analysis to deploy at most six robots in the environment. To illustrate, suppose that two robots are available. Then they are deployed to cells k_{19} and k_6, since these cells command the highest Shapley values. In particular, suppose that one of these robots responds to an event. Then the other robot, if it is available, is dispatched to cell k_{19}. This is the case unless another robot is already responding to an event in that region. In that situation, cell k_6 becomes the next best deployment location. For this example, the Shapley solution to the supergame can be used to deploy at most six free robots as a function of the destinations of the busy robots.

One computationally compelling aspect of SVR is that the analysis need be performed only once each time the demand likelihood distribution changes. These distributions can be easily accommodated by observing event occurrences in an environment over a period of time. Thus a feature of SVR is its adaptability, and the adaptation can be performed off-line. Further, the solution is global in that the rank ordering obtained from the supergame applies to regions that jointly cover the whole environment.

Figures 4.10 through 4.12 (reflecting the data in Tables 4.1 through 4.3) compare the performance of SVR with that of NRR for robot team sizes of 2, 3, and 4, respectively, as a function of the arrival rate λ. As expected, we see that the team is saturated at higher arrival rates as the team size is increased. Saturation, as mentioned before, refers to the arrival rate exceeding the service rate of the team. Another phenomenon apparent from these three graphs is the divergence in performance of the various rules with increasing team size.

As more robots are available for deployment, demand-anticipating rules diverge more acutely from the NRR rule. As the team size increases from 2 to 3 (Figures 4.10 and 4.11), the SVR approach outperforms NRR to a greater degree for the case of 3 robots.

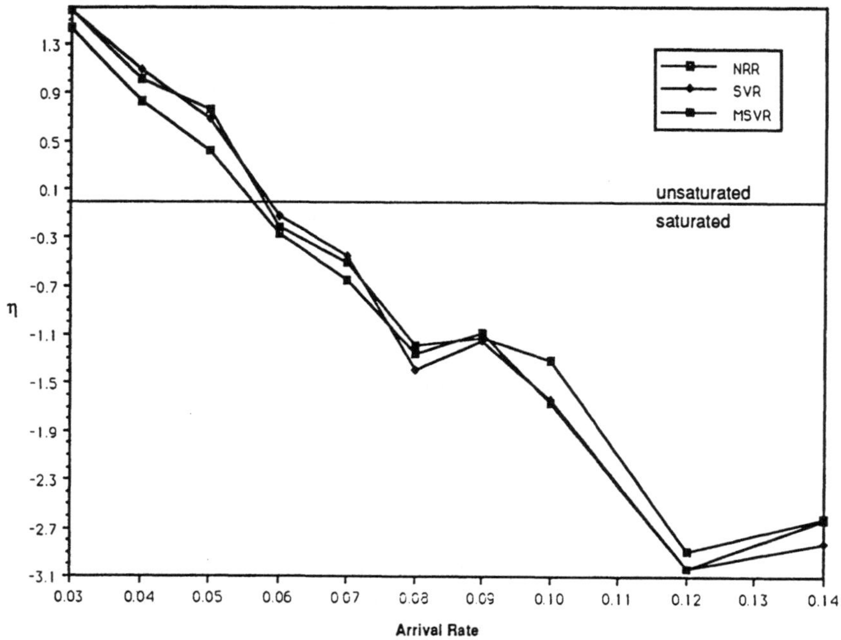

Figure 4.10 Performance of NRR, SVR, and MSVR deployment rules for the 2 robot team.

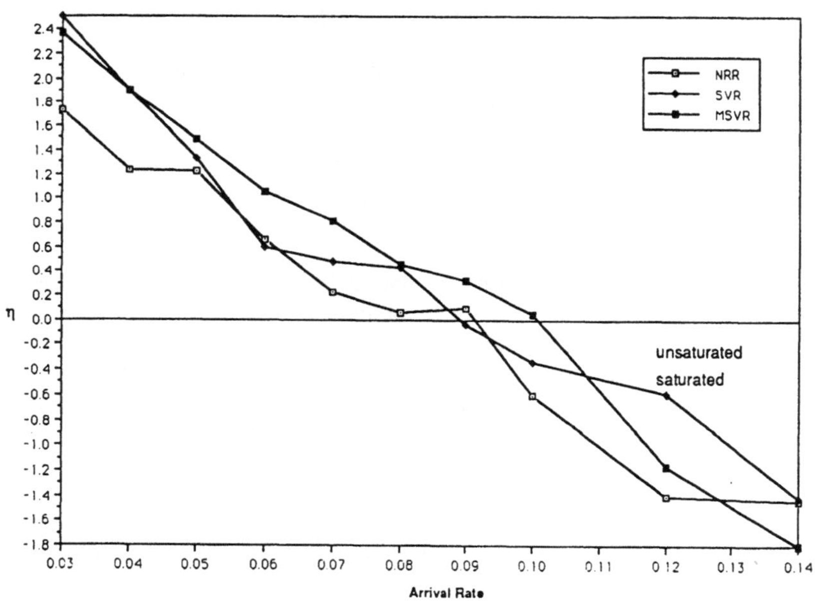

Figure 4.11 Performance of NRR, SVR, and MSVR deployment rules for the 3 robot team.

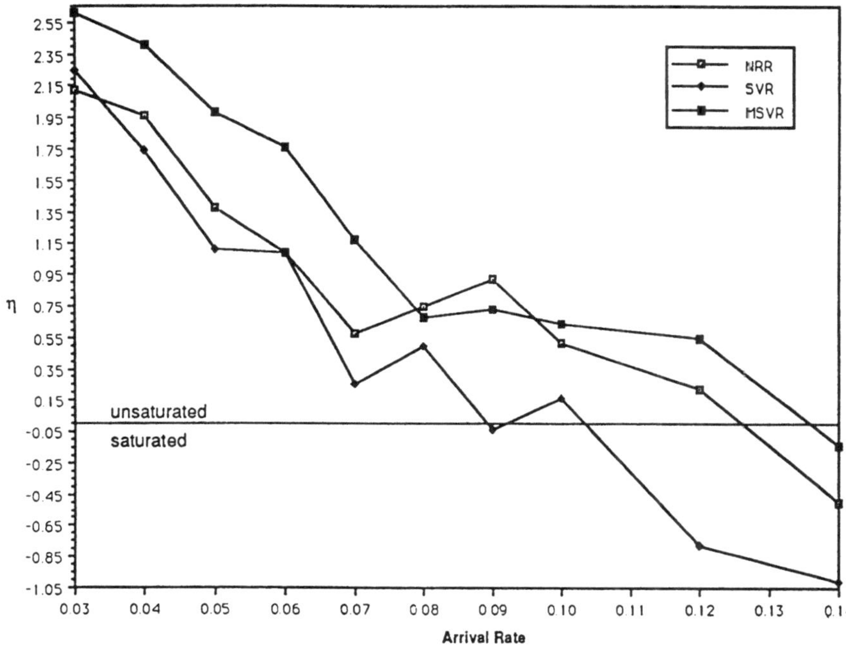

Figure 4.12 Performance of NRR, SVR, and MSVR deployment rules for the 4 robot team.

Table 4.1 Results for the 2 robot team at various event arrival rates (λ). LRT denotes Longest Response Time.

	NRR			SVR			MSVR		
λ	μ	LRT	σ^2	μ	LRT	σ^2	μ	LRT	σ^2
0.03	13.80	54	99.28	11.42	48	87.18	11.05	44	89.89
0.04	15.70	58	147.63	12.41	50	103.54	13.06	57	119.31
0.05	17.65	62	172.43	14.82	48	114.07	12.98	54	107.69
0.06	23.29	76	367.58	22.18	74	275.30	21.54	73	350.57
0.07	29.04	85	453.39	23.52	81	347.83	23.06	89	400.32
0.08	47.13	106	741.57	58.66	111	810.07	47.09	94	638.14
0.09	36.71	105	457.30	32.84	113	591.19	34.97	113	526.89
0.10	60.16	131	765.57	57.96	128	760.64	40.40	86	503.41
0.12	157.80	260	4000.59	145.06	266	4323.84	119.41	239	3716.91
0.14	78.11	185	1971.94	98.97	178	2480.62	78.94	163	1914.50

Table 4.2 Results for the 3 robot team at various event arrival rates (λ). LRT denotes Longest Response Time.

	NRR			SVR			MSVR		
λ	μ	LRT	σ^2	μ	LRT	σ^2	μ	LRT	σ^2
0.03	10.32	38	66.58	6.16	23	19.19	6.94	25	22.83
0.04	10.26	55	88.98	6.72	32	29.54	7.30	28	27.00
0.05	8.85	38	54.96	7.81	41	48.96	7.74	35	34.22
0.06	11.67	49	87.43	12.12	43	95.84	8.90	43	46.09
0.07	12.40	47	141.64	10.49	48	92.92	9.09	34	48.97
0.08	14.42	45	119.92	10.58	36	69.20	10.66	39	64.00
0.09	12.39	36	90.26	13.27	52	113.84	9.98	36	65.43
0.10	19.12	73	208.93	13.53	57	162.09	10.34	42	86.71
0.12	32.61	91	455.68	15.39	52	147.06	21.11	85	397.96
0.14	26.16	84	380.02	25.76	71	396.80	32.41	95	700.14

Table 4.3 Results for the 4 robot team at various event arrival rates (λ). LRT denotes Longest Response Time.

	NRR			SVR			MSVR		
λ	μ	LRT	σ^2	μ	LRT	σ^2	μ	LRT	σ^2
0.03	6.99	40	40.07	6.20	35	33.78	5.66	26	16.13
0.04	6.75	27	25.38	6.68	48	42.08	5.56	14	10.94
0.05	7.19	34	46.48	7.60	51	81.39	5.99	23	13.93
0.06	7.92	34	47.29	7.05	37	53.16	6.36	16	12.59
0.07	8.95	49	86.05	11.06	75	146.08	6.64	41	29.32
0.08	7.99	36	43.51	8.62	54	72.38	8.03	38	50.98
0.09	6.72	24	24.22	11.52	45	128.27	7.36	32	34.52
0.10	8.15	30	37.38	9.65	42	70.98	6.82	27	33.87
0.12	8.27	36	41.99	16.16	54	212.28	7.16	28	23.17
0.14	12.27	40	95.51	15.61	77	239.90	9.64	33	52.44

However, as we further increase the team size to 4 (Figure 4.12), SVR fares worse than NRR at almost all arrival rates. An increase in team size shifts performance levels up for both rules. But what accounts for the trend reversal between the 3-robot and 4-robot teams?

The reason for this reversal becomes clear in watching the simulations unfold for the 3-robot and 4-robot teams. With large team sizes, SVR generates a considerable amount of traffic as it deploys free robots in anticipation of next event locations. Under even low system loading conditions (small value of λ) this increased traffic causes delays for robots attempting to respond to events: they must avoid fellow robots being deployed to anticipatory locations as well as humans. NRR causes no such delays, since all free robots are stationary at their last dispatched locations with this rule. As the average time between events decreases – corresponding to increasing λ – the traffic congestion caused by anticipatory deployment via the SVR rule becomes more critical. Figure 4.12 reflects this in the way that the performance of SVR falls more precipitously than that of NRR as λ increases.

Overall, we find that anticipation in the form of SVR increases performance of the robotic team. However, it also causes increased traffic, thereby reversing the rise in performance with team size. NRR has an advantage in these situations, since robots remain stationary when free, thereby avoiding traffic congestion in the environment. It would be preferable to retain the increase in performance due to demand anticipation while reducing the traffic congestion caused by the anticipatory deployment of free robots.

Modified Shapley Value Rule (MSVR). The third curve on each of the charts in Figures 4.10 through 4.12 is labeled MSVR. It is no accident that MSVR fares better than either the SVR or NRR. In fact, MSVR originated from observing the results of SVR and NRR, and evaluating their respective advantages and liabilities.

MSVR employs the results from the subgames played among the regions of each superregion. When a robot becomes free (after having arrived at the location of the previous event), it is sent to the capital cell of the region with the largest value of φ from the subgame associated with the superregion containing the robot. If either another robot is responding to an event there or another free robot is already stationed there, the adjacent superregions are considered. The adjacent superregions are rank ordered in terms of the Shapley values of their winning regions (a winning region being one which has the highest value of φ in the subgame played among the regions of a superregion). The robot then seeks the capital cell of the region ranked highest among those that do not have other robots deployed there. Thus anticipatory coverage is exercised but robot deployment is kept local. Traffic congestion falls sharply, since anticipating robots limit the use of the elevator, which is the bottleneck resource. Local anticipation with MSVR decreases the use of the elevator by anticipating free robots, thus allowing more freedom to robots that are actually responding to events.

Another nice feature of MSVR is that it adapts to trends in event locations. That is, if many events occur in sequence in some superregion, MSVR will keep robots close to that superregion, thereby reducing subsequent delays in response. NRR has the same property; however, NRR cannot adapt to the situation where events space themselves out over the entire environment. SVR and MSVR can adapt to this latter case by anticipating event locations over the entire environment. SVR, however, over-anticipates and cannot adapt to local event sequences. Thus MSVR, having both of these desirable properties, outperforms both SVR and NRR. The key element that makes MSVR a better deployment rule than either SVR or NRR is that it has both anticipatory and adaptive properties.

One final point remains to be made in connection with the simulation results in Figures 4.10 through 4.12. This is the nonmonotonicity of the performance curves. In all three figures we see that at arrival rates between 0.08 and 0.1, the curves undulate, arresting or even reversing their previous downward trends. This behavior can be explained as follows. Certain combinations of traffic congestion conditions, system loading values, and event location histories reduce response delays, thereby artificially boosting the performance of each deployment rule. This is especially true for NRR and MSVR, which capitalize on event location histories which exhibit long sequences of proximal events.

Conclusion

The practical consequences of coordinative principles were examined through a case study in robotic deployment. Deployment rules based on game theoretic concepts were developed and evaluated against a non-anticipatory deterministic rule (NRR).

For algorithmic simplicity, the agents were considered as the regions in the environment, and the robots as resources to be allocated. The Shapley values associated with the resulting game can be used to construct deployment rules: in particular, SVR and MSVR. The last rule (MSVR) was developed after an evaluation of the relative advantages of the former two strategies. MSVR exhibits both locally and globally adaptive properties. The locally adaptive characteristics of NRR and the globally anticipatory properties of SVR were combined in this new rule which was found to outperform the other two for various team sizes and most of the system loading conditions.

In this way, the principles of coordination and adaptation examined in previous chapters have been given practical expression. The encouraging results reflect the promise of game-theoretic strategies and their extensions to other domains of coordination among autonomous agents.

References

Alexandris, N.A., G.C. Barney, and C.J. Harris, "Multi-Car Lift Systems Analysis and Design", *Applied Mathematical Modelling*, Vol. 3, pp. 269-274, 1979a.

Alexandris, N.A., G.C. Barney, and C.J. Harris, "Derivation of the Mean Highest Reversal Floor and Expected Number of Stops in Lift Systems", *Applied Mathematical Modelling*, Vol. 3, pp. 275-279, 1979b.

Alexandris, N.A., G.C. Barney, and C.J. Harris, "Evaluation of the Handling Capacity of Multi-Car Lift Systems", *Applied Mathematical Modelling*, Vol. 5, pp. 49-52, 1981.

Aumann, R.J., "Subjectivity and Correlation in Randomized Strategies", *Journal of Mathematical Economics*, 1, pp. 67-96, 1974.

Aumann, R.J., "Correlated Equilibrium as an Expression of Bayesian Rationality", *Econometrica*, Vol. 55, No. 1, pp. 1-18, 1987.

Cheng, T.C.E., "A Simulation Study of Automated Guided Vehicle Dispatching", *Robotics and Computer Integrated Manufacturing*, 3, pp. 335-338, 1987.

Denardo, E.V., and B.L. Fox, "Shortest-Route Methods: 1. Reaching, Pruning and Buckets." *Operations Research*, Vol. 27, pp. 161-186, 1979.

Dreyfus, S.E., "An Appraisal of Some Shortest-Path Algorithms." *Operations Research*, Vol. 17, pp.395-412, 1969.

Egbelu, P.J., "Pull versus Push Strategy for Automated Guided Vechile Load Movement in a Batch Manufacturing System", *Journal of Manufacturing Systems*, 6, pp. 209-220, 1986.

Egbelu, P.J. and J.M.A. Tanchoco, "Characterizations of AGV Dispatching Rules", *Automated Guided Vechile Systems*, pp. 125-141, 1987.

Egilmez, K. and S. H. Kim. "Deployment of Robotic Agents in Uncertain Environments: Game Theoretic Rules and Simulation Studies". *Artif. Intel. for Engineering Design, Analysis and Manufacturing*, v. 6(1), 1992, pp. 1-17.

Kim, S.H., "Coordination of Multiagent Systems through Explicit Valuation of Action.", *Robotics and Computer-Integrated Manufacturing*, v. 8(4), 1991, pp. 265-291.

Nof, S.Y., and V.N. Rajan, "Planning Collaborative Automation for Flexible Assembly", *Technical Paper MS90-841: 11th International Conference on Assembly Automation*, Dearborn, MI, Nov. 11-14, 1990.

Nof, S.Y., "Game Theoretic Models for Planning Cooperative Robotic Work", *Proc. of the NSF Design and Manufacturing Systems Conf.*, Austin, TX, Jan. 1991.

Rapoport, A., *N-Person Game Theory: Concepts and Applications*, The University of Michigan Press, Ann Arbor, MI, 1970.

Tsai, J.J.-P., M. Metea, and J. Cesarone, "A Knowledge-Based Navigation Scheme for Autonomous Land Vehicles", *Applied Artificial Intelligence*, Vol. 4, No. 1, pp. 1-14, 1990.

Conclusion 5

A hallmark of intelligent systems is the ability to learn. Despite the rich set of examples of learning in the biological realm, synthetic systems tend to learn poorly, if at all. A better understanding of the nature and mechanisms of learning should enable designers to better emulate the abilities of natural systems.

When intelligent systems act in concert, their coordination is often of greater consequence for their collective objective than are their individual capabilities. To date, much work in coordination has relied on centralized strategies. However, the limitations of centralized control underscore the need for decentralized policies. Effective coordination becomes more paramount as systems of autonomous agents assume a greater role in the engineered world around us.

This concluding part first summarizes the results of the preceding chapters. The subsequent discussion presents a number of implications for the future.

Summary

We began in Chapter 1 with a general algebraic model for learning systems. This model was interpreted for systems incorporating knowledge in diverse forms. The knowledge might be stored in a high-level form such as production rules, or in a low-level format such as the link weights in neural networks.

The second chapter examined neural nets of both the natural and artificial kinds. A comparative study highlighted the similarities and differences between these types of networks.

At the structural level, the human brain exhibits a richness of organization: many types of cells process signals in both discrete and analog forms. Moreover, the cells themselves are organized into modules for specialized processing, and within modules into distinct layers ranging from elementary signal transduction to the recognition of complex features, and even to the formation of relationships among abstract concepts.

In the realm of operations, learning can occur with great speed in a biological system. Often a single exposure to a stimulus is sufficient for comprehension and retention of the salient aspects for future reference.

The celerity of learning in the biological realm contrasts sharply with artificial devices, whether the latter are organized in the form of networks or otherwise. To date, artificial neural nets have exhibited only simple organizational forms. Often only a single type of cell is connected in elementary ways to other units in no more than a handful of layers. The stratagem of increasing the structural sophistication of artificial systems leaves much room for organizational elaboration and performance enhancement.

An intelligent system often consists of relatively independent agents in interaction. The overall performance of such systems may depend more on the ability of the group to coordinate behavior than on the portfolio of individual capabilities.

Emergent properties often arise when autonomous agents interact. One related problem lies in paralysis through deadlock. Paralysis can occur even in the simplest collective comprising only two agents faced with two choices, or three agents with three choices. The opportunities for partial or complete paralysis are even greater in larger systems.

An effective way to coordinate activity is to assign a value to each potential action. The valuation of behavior in terms of benefits and costs corresponds to an economic perspective of decision making. This is the province of game theory and related methodologies.

The richness of the behavior space for a set of agents rises dramatically with their number. Even so, the concept of strategic equivalence underscores the similarities among large classes of decision making situations; the strategic consequences for a canonical situation applies without exception to all the games in the same class.

A number of solution methods are available for ascertaining the relative value of each agent in a coalition. Each of these methods has its strengths and limitations.

One effective technique is the Shapley approach, which is tantamount to assigning a relative worth to an agent based on his expected contribution to the synergism of the group. A key feature of the Shapley value lies in existence and uniqueness properties: it is guaranteed to exist and to possess a unique value.

A number of strategies for multiagent coordination were examined in Chapter 4. The case study involved a robotic deployment scenario. The associated simulation studies verified the superiority of game theoretic approaches to a straightforward decision rule.

The case study also incorporated adaptation in connection with the use of deployment strategies. This application demonstrates the harmonious interaction between learning and coordination. A promising direction for further work lies in a comprehensive examination of these phenomena—namely, the impact of coordination on the effectiveness of learning strategies and vice versa.

Implications

The results to date offer cause for optimism in the study and construction of intelligent systems. At the structural level, the simplicity of engineered systems in the past may well have been responsible for their limited performance. Increasing the sophistication at the hardware level alone might yield dramatic improvements in performance.

Even in the closing years of the twentieth century, synthetic devices exhibit no more than about 10^5 processing units. Since the human brain has perhaps 10^{11} neurons, there is a million-fold gap along this parameter alone. The emergent nature of systems highlights the fact that an increase in quantity can lead to qualitative changes as well. For instance, information is routinely transmitted through sound waves characterized by pitch, amplitude, phase, and direction. Yet the waves are superimposed on discrete molecules that collectively travel in all directions with unpredictable velocities.

Another cause for optimism lies in the deliberate nature of engineering design as opposed to the passive nature of evolution. Organs and organisms depend on random genetic mutations, and are composed of substances available in the prevailing environment. Artificial devices, on the other hand, can make use of synthetic substances which are fashioned toward deliberate ends.

As a result, we have semiconductor devices which can change states within picoseconds. In contrast, the ionic channels of carbon-based, organic neurons require milliseconds to alter states. This billion-fold disparity in switching speed can only work in favor of artificial devices as they are fashioned to exhibit rich behavior spaces in real-time.

The gap in processing speed lies at a low level. The potential difference in performance due to system architecture should be even more accentuated. Our collective ability to organize goods, services, and individuals into cities, economies, and societies provides a rich knowledge base for developing versatile systems.

We are still unable to fully comprehend the nature and operation of these large-scale, megalevel institutions. Even so, we can still manage and control them, with some success, to serve our ends. The lessons from the fields of economics, sociology, governance, and other arenas can assist us in the development of intelligent systems. Some directions along these lines were discussed in Chapters 3 and 4.

Our increasing understanding of the life sciences, management arts, and social studies augur well for the development of sophisticated systems. The principles of learning behavior and coordinated interaction arising from engineered artifacts may, in turn, offer lessons for guiding the evolution of biological systems and societal institutions.

Dynamic Models in Statistical Physics — Appendix

In an uncertain world, learning is an inherently stochastic process. As a result, one way to model the interplay between individual behavior and collective performance lies with the techniques of stochastic dynamics. This approach first originated in the physics of thermal phenomena. The methodology, however, is applicable to other arenas involving large collections of objects. One such field is that of neural networks.

This appendix presents the basis for dynamic models in statistical physics. The concepts serve as the foundation for the statistical models discussed in Chapter 2.

Emergence

The ability to transcend one's initial capabilities is a compelling attribute of a complex system. Among these attributes, a key characteristic is the ability to learn. Growth in knowledge or performance is a remarkable phenomenon rivaled by another, more general characteristic. This is the property of ***emergence***: the tendency for a complex system to display global characteristics which are lacking at the level of individual components that comprise the system. Examples of emergence are found in all spheres of natural and human activity, as exemplified by the flowering of life out of lifeless components or the onset of instability in a system composed of stable elements (Kim, 1990 etc.).

This appendix examines the interplay of learning and emergence in complex systems. In this context, a *complex* system refers to one composed of many components. Further, the components are sufficiently numerous that the effect of any single element is marginal or insignificant. The components exert their effects on the environment through the action of the entirety. In this activity, the components may or may not interact with each other according to some prescribed rule. Their actions may therefore be coordinated or purely random. Whatever the nature of their interactions, the behavior of the system can be usefully described in terms of global parameters. These may relate to the average impact of individual activity, as exemplified by the temperature due to gas particles in a vessel; or their combined effect, such as total energy or variations in characteristics such as sound waves.

Statistical mechanics

Statistical mechanics involves the study of systems which consist of numerous particles. These particles may number in the hundreds or quintillions; but they are of a single variety or at most a handful of comparable species.

Dual State System

We begin with the simple case of particles which can take one of two states. An example lies in the form of a hydrogen atom, which might be in the ground state or excited state.

Suppose there are N particles, of which N_1 are in the ground state and N_2 in the excited state. Obviously, the variables are constrained by the relation $N_1 + N_2 = N$.

Consider the question "In how many ways can N_2 particles be in the excited state?" The answer is simply the binomial coefficient given by

$$\binom{N}{N_2} \equiv \frac{N!}{N_2!(N-N_2)!} = \frac{N!}{N_1!N_2!}$$

The pair of values (N_1, N_2) is called a *configuration* of the system. Since there are only two states, we have $N_2 = N - N_1$. Thus the configuration is parametrized by a single degree of freedom. We may view the configuration as being defined by $(N_1, N - N_1)$ or as $(N - N_2, N_2)$; the former uses the number N_1 of particles in the ground state, while the latter uses the number N_2 of objects in the excited state.

Note that each configuration corresponds to a distinct level of energy for the system. To illustrate, suppose that the energy required to drive a single particle from the ground to excited state is 10 electron volts. The configuration $N_2 = 0$ corresponds to zero energy for the system, while $N_2 = 3$ corresponds to $3(10) = 30$ e.v.

Each distinctive pattern of states for the system is called a *microstate*. Let "+" denote a particle in the excited state, and "–" one in the ground state. Then for the case of N = 4 objects, the pattern (+, –, –, –) indicates one microstate, while (–, +, –, –) denotes a different microstate. Both these microstates, however, belong to the same configuration; namely, one excited particle in the system, for which $N_2 = 1$.

The number of microstates which belong to the same configuration is called the *degeneracy* of the configuration. For the case of $N_2 = 1$, the degeneracy is 4, since the allowable microstates are (+, –, –, –), (–, +, –, –), (–, –, +, –), and (–, –, –, +). Figure A.1 presents the complete space of microstates, configurations, and associated degeneracies for the case of a 4-particle system.

Configuration N_1	Configuration N_2	Degeneracy	Microstates			
4	0	1	− − − −			
3	1	4	+ − − −	− + − −	− − + −	− − − +
2	2	6	+ + − −	+ − + −	+ − − +	
			− + + −	− + − +	− − + +	
1	3	4	+ + + −	+ + − +	+ − + +	− + + +
0	4	1	+ + + +			

Figure A.1 Possible patterns for a dual-state system containing four particles.

Multi-state System

We now turn to the general case of multiple states for each particle in a system. Suppose that each particle has the same mass m. Let x_i, y_i, and z_i be the coordinates of the i th particle; the component of the velocity along each axis is denoted by v_{ix}, v_{iy}, and v_{iz} respectively. The speed of the particle is given by $v_i = [v_{ix}^2 + v_{iy}^2 + v_{iz}^2]^{1/2}$. The translational energy for the i th particle is

$$\tau_i = \frac{mv_i^2}{2}$$

The potential energy φ_i of particle i is due to its location in a conservative force field. This field may be due to a factor such as an electrical, magnetic, or gravitational force. In the case of a gravitational field, for instance, the potential energy is $\varphi_i = mgz_i$ where g is the gravitational acceleration and z_i is the height of particle i above some appropriate reference level.

The particle may also possess an internal energy ζ_i. This may be due to factors such as rotation, vibration, or electrical excitation.

The nominal energy ε_i of particle i is given by the sum of the translational, potential, and internal energies:

$$\varepsilon_i = \tau_i + \varphi_i + \zeta_i$$

For any specific application, each of these energy components may be large, small, or irrelevant.

From the field of quantum mechanics, we know that each of variables τ_i, φ_i, and ζ_i assumes discrete values. Consequently, the energy level ε_i will also be discrete.

Suppose that the system consists of N particles. Then the overall energy of the system will be the sum of the individual energies:

$$E = \sum_{i=1}^{N} \varepsilon_i$$

Since the microlevel energy ε_i is a discrete parameter, so is the macroscopic energy level E. In other words, the global energy E can assume values in the set $\{E_1, E_2, ..., E_J\}$ where J is the total number of discrete energy levels.

Let N_j denote the number of distinct microstates which correspond to E_j; that is, N_j is the *degeneracy* of the *energy* level E_j. The set of degeneracy values $(N_1, N_2, ..., N_J)$ is called a *configuration* of the system.

Of course, the degeneracies N_j must add up to the total number of particles:

$$N = \sum_{j=1}^{J} N_j \qquad \text{(Eq. 1)}$$

Hence the configuration $(N_1, ..., N_J)$ is parametrized by $J — 1$ degrees of freedom.

The total energy E of the system can be stated in terms of the degeneracy levels N_j. In particular, we have

$$E = \sum_{j=1}^{J} N_j E_j \qquad \text{(Eq. 2)}$$

We can also speak of the *degeneracy of a configuration* $(N_1, ..., N_J)$: the degeneracy of the number of microstates which corresponds to the configuration. This value is given by the multinomial coefficient:

$$W(N_1, ..., N_J) = \frac{N!}{\prod_{j=1}^{J} N_j!} \qquad \text{(Eq. 3)}$$

Modal Configuration

Consider an isolated system of N particles in a volume V. Since the system is isolated, the global energy E is constant. We would like to ascertain the configuration $(N_1, ..., N_J)$ that is most likely to be observed.

To that end, we invoke a principle of indifference: we assume that each microstate is as likely to occur as any other. Under this assumption, the most probable configuration is one that exhibits the highest degeneracy $W(N_1, ..., N_J)$.

To determine this configuration, we can seek the maximum value of W subject to the conservation formula in Eq. 1 and the energy constraint in Eq. 2. Since W increases monotonically with ln W, it is sufficient to maximize the latter quantity. From Eq. 3 we obtain

$$\ln W = \ln N! - \sum_{j=1}^{J} \ln N_j!$$

At this point we invoke Stirling's formula

$$\ln X! \approx X \ln X - X$$

which is increasingly accurate for large values of X. As a result, ln W can be written as

$$\begin{aligned} \ln W &= (N \ln N - N) - \Big[\sum_{j=1}^{J} N_j \ln N_j - \sum_{j=1}^{J} N_j\Big] \\ &= N \ln N - \sum_{j=1}^{J} N_j \ln N_j \end{aligned}$$

The second equality depends on the constraint in Eq. 1.

We can form the Lagrangian of ln W using Eqs. 1 and 2 to obtain

$$L = N \ln N - \sum_{j=1}^{J} N_j\, ln\, N_j - \alpha(\sum_{j=1}^{J} N_j - N) - \beta(\sum_{j=1}^{J} N_j E_j - N)$$

We then take the partial derivative $dL\,/\,dN_j$ and set it to zero. After some rearrangement, we have

$$N_j = e^{-\alpha-1}\, e^{-\beta E_j}$$

Injecting this relation into Eq. 1 yields

$$N = \sum_{j=1}^{J} N_j = e^{-\alpha-1} \sum_{j=1}^{J} e^{-\beta E_j}$$

Hence

$$e^{-\alpha-1} = \frac{N}{\sum_{j=1}^{J} e^{-\beta E_j}} = \frac{N}{Z(\beta)}$$

where we have defined $Z(\beta) \equiv \Sigma\, e^{-\beta E_j}$. The quantity $Z(\beta)$ is called the *partition function*, for reasons to be explained shortly.

Combining the results from the last two paragraphs, we have

$$N_j = \frac{N}{Z(\beta)} e^{-\beta E_j} \qquad \text{(Eq. 4)}$$

For a given set of energy levels, both N and $Z(\beta)$ are fixed. Hence the degeneracy N_j decreases exponentially with the energy E_j; the rate of decrease depends solely on the parameter β.

We can also write Eq. 4 as

$$p_j = \frac{N_j}{N} = \frac{e^{-\beta E_j}}{Z(\beta)} \qquad \text{(Eq. 5)}$$

Since N_j/N is the fraction of the particles in energy level E_j, the quantity p_j denotes the probability that a particle chosen at random will inhabit that energy level. The probability p_j is given by the ratio of the term $e^{-\beta E_j}$ to the sum of all such terms, namely $Z(\beta)$.

Figure A.2 depicts the probability p_j as a function of the energy level E_j. The function drops sharply for large values of β, and gently for small values of the parameter.

As indicated earlier, $Z(\beta)$ is called the partition function of the system. The function is defined as

$$Z(\beta) = e^{-\beta E_1} + \cdots + e^{-\beta E_j} + \cdots + e^{-\beta E_J}$$

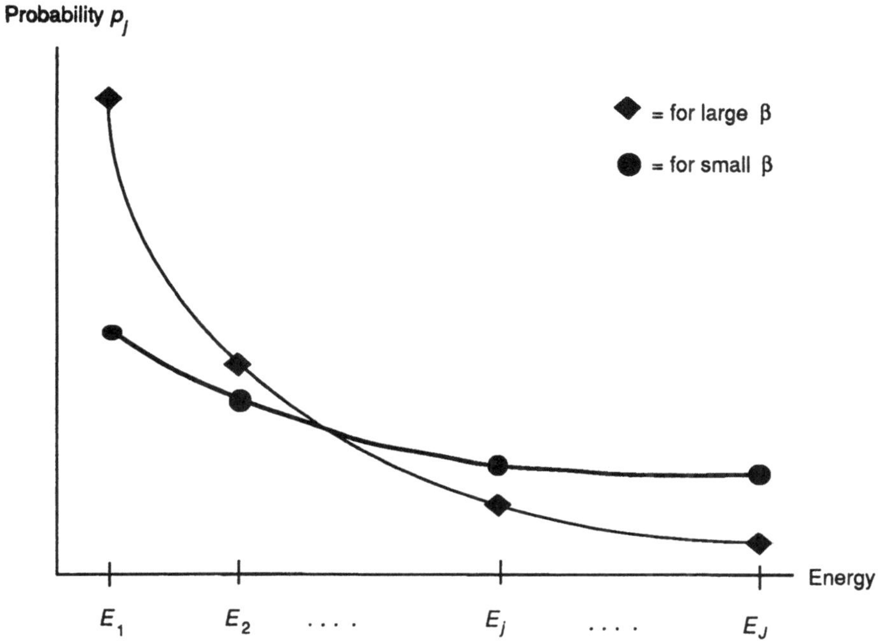

Figure A.2 The probability p_j that a random particle has energy E_j, for different values of the parameter β.

From Eq. 5, we see that the jth term in $Z(\beta)$ gives the likelihood that a particle chosen at random has energy E_j. In other words, the terms in $Z(\beta)$ partition the probability of finding particles among all the energy levels.

Temperature Parameter

Our next task is to obtain a physical interpretation for the parameter β in the partition function $Z(\beta)$. From Eq. 4, we know that β must have units of $(\text{energy})^{-1}$, since the product βE_j appears as an exponent and must be dimensionless. We can be more specific about the nature of β by considering some specific physical scenarios.

Consider once again a system of N particles isolated in a volume V. The volume is partitioned into J regions $\{\tilde{r}_1, \ldots, \tilde{r}_J\}$, as indicated in Figure A.3. Each region has volume $\tilde{r}_j$ and contains N_j particles. We assume the region is large enough so that N_j is often much greater than 1. On the other hand, the region must be small enough that each of the N_j particles has energy approximately equal to a common value E_j.

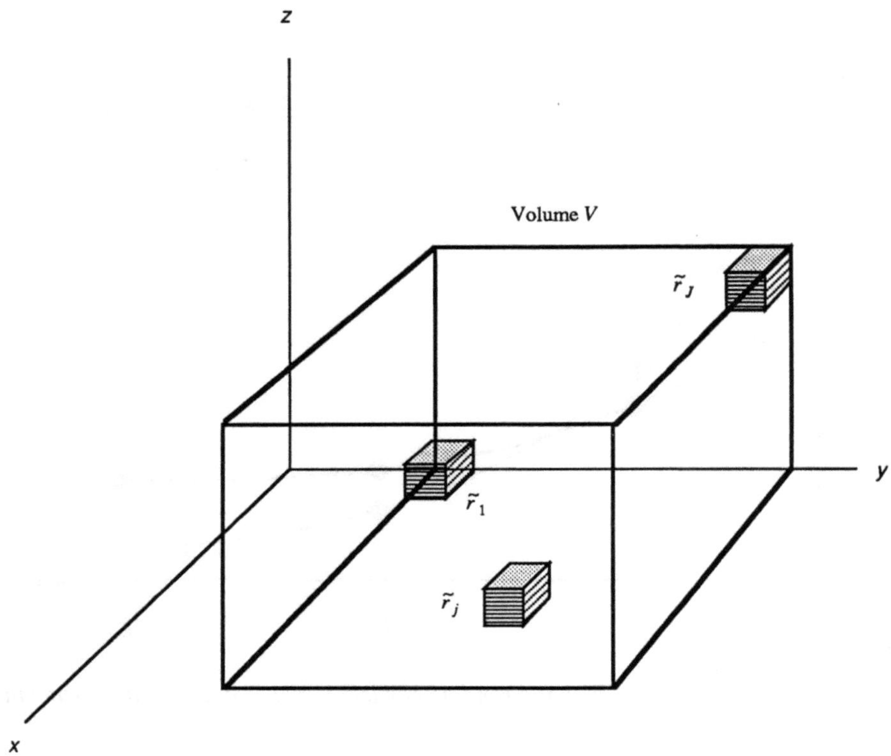

Figure A.3 An isolated system of volume V, partitioned into regions $\tilde{r}_1, \ldots, \tilde{r}_J$

We recall that the energy ε_i of a particle can be partitioned into the translational, potential, and internal components:

$$\varepsilon_i = \tau_i + \varphi_i + \zeta_i$$

By summing over all N_j particles in region $\tilde{r}_j$, we obtain

$$E_j = E_j(\tau) + E_j(\varphi) + E_j(\zeta)$$

The three terms on the right correspond respectively to the terms in the preceding equation. For instance,

$$E_j(\tau) \equiv \sum_{i=1}^{N_j} \tau_i$$

From Eq. 4, we can write

$$N_j = \frac{N}{Z(\beta)} e^{-\beta E_j} = \frac{N}{Z(\beta)} e^{-\beta\,[E_j(\tau) + E_j(\varphi) + E_j(\zeta)]}$$

$$= \frac{N\, e^{-\beta\,[E_j(\tau) + E_j(\zeta)]}}{Z(\beta)} e^{-\beta E_j(\varphi)}$$

Suppose that the translational and internal energies of the particles in all J regions are approximately equal. Then $E_j(\tau)$ and $E_j(\zeta)$ are fixed, and the preceding equation can be written as

$$N_j = A(\beta)\, e^{-\beta \varphi_j} \qquad \text{(Eq. 6)}$$

Here we have made the substitutions $\varphi_i \equiv E_j(\varphi)$ and $A(\beta) = N\exp[-\beta(E_j(\tau) + E_j(\zeta)]/Z(\beta)$. In Eq. 6 we find the Boltzmann distribution law for particles in a potential field.

To continue with our quest for a physical intepretation for the parameter β, we consider a fluid in a gravitational field. We will derive the consequences due to two perspectives, then unify the results to obtain a formula for β.

We multiply Eq. 6 by the quantity m/r_j, where m is the mass of each particle:

$$\frac{m\, N_j}{r_j} = \frac{m\, A(\beta)}{r_j} e^{-\beta \varphi_j} \qquad \text{(Eq. 7)}$$

The left side of this formula simply represents the density $\rho = \rho(\varphi_j)$ for a fluid in region r_j.

For a gravitational field, the potential energy is given by $\varphi_j(x, y, z) = mgz$. Since φ_j depends only on the height z, we may write $\varphi_j = \varphi_j(z)$. Consequently the density is also a function only of z_j ; that is, $\rho_j = \rho_j(z)$.

Using these results, we may write Eq. 7 in the form

$$\rho_j(z) = \frac{m\, A(\beta)}{r_j} e^{-\beta m g z_j} \qquad \text{(Eq. 8)}$$

At ground level where $z \equiv 0$, we obtain $\rho_j(0) = mA(\beta)/r_j$. As a result, Eq. 8 can now be written as $\rho_j(z) = \rho_j(0) \exp(-\beta m g z_j)$. The only aspect of region r_j which has direct relevance in this formula is its height z. So we may drop the subscript j and simply write

$$\rho(z) = \rho(0)\, e^{-\beta m g z} \qquad \text{(Eq. 9)}$$

This is the standard barometric formula for the increase in atmospheric density as a function of height.

Let us derive this basic relationship from a macroscopic perspective. For one mole of particles, we can write $PV = RT$ where P is the pressure and T the temperature of a gas in a volume V, and $R = 8.314 \times 10^7$ erg/mole °K is the usual gas constant. We can write the gas formula in the following form:

$$P = \frac{RT}{V} = \left(\frac{RT}{m\,N_A}\right)\left(\frac{m\,N_A}{V}\right) = \left(\frac{RT}{m\,N_A}\right)\rho(z)$$

The second equality results from appending the factor mN_A both above and below the dividing line; here $N_A \equiv 6.023 \times 10^{23}$ mole^{-1} is Avogadro's number. The third equality depends on the recognition that mN_A/V is the density ρ of a gas at height z.

Suppose the gas is at equilibrium. Then P is a function of the height z. The hydrostatic pressure gradient is given by

$$\frac{dP}{dz} = -\,g\rho(z)$$

Substituting the previous expression for the pressure P yields

$$\frac{d}{dz}\left[\frac{RT}{m\,N_A}\,\rho(z)\right] = -\,g\rho(z)$$

The temperature T is a constant at equilibrium, and straightforward integration results in

$$\rho(z) = \rho(0)\,e^{-\,mgzN_A/(RT)} \qquad \text{(Eq. 10)}$$

A comparison of Eqs. 9 and 10 yields

$$\beta = \frac{N_A}{RT} = \frac{1}{kT}$$

In the last expression, k is Boltzmann's constant, defined as the gas constant per molecule: $k \equiv R/N_A = 1.381 \times 10^{-16}$ erg/°K. To summarize, the parameter β is inversely proportional to the temperature T and has a scaling factor of $1/k$.

Use of the exponential distribution

The probability distribution defined by Eq. 5 is the basis for the Boltzmann algorithm described in Chapter 2. The parameter β is proportional to some measure of temperature; the role of temperature in simulated annealing was described in that chapter.

Suppose that a weight on a link within a neural network can assume only binary values. Then the probability of its attaining a 1 is given by $\exp(-\beta E_j)/Z(\beta)$ when the value of the input is E_j.

On the other hand, suppose that a weight can assume values on an interval such as [0, 1). Then Eq. 5 can be regarded as the basis for an exponential density function with parameter β:

$$f(E_j) \equiv \beta e^{-\beta E_j}$$

This results from defining $Z(\beta) \equiv 1/\beta$ in Eq. 5. The density function f is defined on the nonnegative real line; that is, the input E_j can assume any value in the interval $[0, \infty)$.

To be of operational consequence, the density function is first transformed into its cumulative distribution function:

$$F(E_j) \equiv \int_0^{E_j} \beta e^{-\beta x} dx = 1 - e^{-\beta E_j}$$

This function rises monotonically with the input E_j, from a low of 0 at $E_j = 0$ toward the asymptote of 1 as $E_j \to \infty$. Given any nonnegative input E_j, the output is some value $F(E_j)$ lying on the unit interval.

Suppose that the possible range of outputs is some interval between any two values a and b rather than [0, 1). A linear transformation of F yields the desired result:

$$\tilde{F}(E_j) \equiv a\,[1-F(E_j)] \;+\; b\,F(E_j)$$

In this way the weight increment can assume values on an arbitrary interval.

The nonnegative restriction on the input E_j can be eliminated in an analogous way. In particular, if E_j can assume values as low as c, where c is negative, then adding $|\,c\,|$ to E_j and using the transformed value as the new input will circumvent the limitation on nonnegative values. In other words, the new variable $E_j' \equiv E_j - c$ will assume values on the interval $[0, \infty)$ when c is a negative constant.

These types of transformations ensure that the exponential distribution can be used even when the input or output takes values on arbitrary intervals.

References

Aitchison, J. and I. R. Dunsmore. *Statistical Prediction Analysis.* Cambridge: Cambridge University Press, 1975.

Andrews, F. C. *Equilibrium Statistical Mechanics,* 2nd ed. NY: Wiley, 1975.

Brout., R. *Phase Transitions.* NY: Benjamin, 1965.

Doremus, R. H. *Rates of Phase Transformations.* NY: Academic, 1985.

Garrido, L., ed. *Statistical Mechanics of Neural Networks.* NY: Springer-Verlag, 1990.

Haken, H. *Advanced Synergetics.* Berlin: Springer-Verlag, 1983.

Haken, H. *Synergetics,* 3rd ed. Berlin: Springer-Verlag, 1983.

Holden, A. V. *Models of the Stochastic Activity of Neurons.* Berlin: Springer-Verlag, 1976.

Jaynes, E. T. "Information Theory and Statistical Mechanics." *Physical Review, 106(4),* 1957, pp. 620-30.

Kim, S. H. *Designing Intelligence: A Framework for Smart Systems.* New York: Oxford Univ. Press, 1990.

Knuth, E. L. *Introduction to Statistical Thermodynamics.* NY: McGraw-Hill, 1966.

Livi, R., S. Ruffo, S. Ciliberto, and M. Buiatti, eds. *Workshop on Chaos and Complexity.* Teaneck, NJ: World Scientific, 1988.

Lu, S. C.-Y. and D. K. Tcheng. "Building Layered Models to Support Engineering Decision Making: A Machine Learning Approach." *J. of Engineering for Industry, 113(1),* 1991, pp. 1-9.

Mezard, M., G. Parisi, and M. A. Virasoro. *Spin Glass Theory and Beyond.* Singapore: World Scientific, 1987.

Nicolis, G., and I. Prigogine. *Exploring Complexity.* NY: Freeman, 1989.

Nicolis, G. and I. Prigogine. *Self-Organization in Nonequilibrium Systems.* NY: Wiley, 1977.

Prigogine, I. *Introduction to Thermodynamics of Irreversible Processes.* NY: Wiley, 1961.

Rumelhart, D. E., G. E. Hinton, and R. J. Williams. "Learning Internal Representations by Error Propagation." In Rumelhart, McClelland, et al. (1986), pp. 318-362.

Schrödinger, E. *Statistical Thermodynamics.* NY: Cambridge University Press, 1946.

Sejnowski, T. J. and C. R. Rosenberg. "Parallel Networks that Learn to Pronounce English Text." *Complex Systems*, v. 1, 1987, pp. 145-168.

Skilling, J., ed. *Maximum Entropy and Bayesian Methods.* Boston: Kluwer, 1989.

Tribus, M. *Thermostatistics and Thermodynamics.* NY: Van Nostrand, 1961.

Tribus, M., and R. D. Levine, eds. *The Maximum Entropy Formalism.* Cambridge, MA: MIT Press, 1979.

Tsetlin, M. L. "On Behavior of Finite Automata in Random Media." *Automation and Remote Control,* v. 22, 1961, pp. 1345-1354.

Winfree, A. T. *The Geometry of Biological Time.* NY: Springer-Verlag, 1980.

Index

MIX
Papier aus verantwortungsvollen Quellen
Paper from responsible sources
FSC® C105338

If you have any concerns about our products,
you can contact us on
ProductSafety@springernature.com

In case Publisher is established outside the EU,
the EU authorized representative is:
Springer Nature Customer Service Center GmbH
Europaplatz 3, 69115 Heidelberg, Germany

Printed by Libri Plureos GmbH
in Hamburg, Germany